MATHEMATICS
for New Speakers of English

A Supplemental Text for Beginning ELD Students

MATHEMATICS
for New Speakers of English

Three Watson
Irvine, CA 92618-2767

Website: www.sdlback.com

Copyright ©2005 by Saddleback Publishing, Inc. All rights reserved. No part of this book may be reproduced in any form or by any means, electronic or mechanical, including photocopying, recording, or by any information storage and retrieval system, without the written permission of the publisher.

ISBN 1-56254-646-5

Printed in the United States of America
10 09 08 07 06 9 8 7 6 5 4 3 2 1

Table of Contents

Chapter 1 Names of Numbers

1.1	Names of Numbers	3
1.5	More Names of Numbers	6
1.9	Even More Names of Numbers	9
1.13	Number Periods	12
1.18	Names of Decimal Fractions	16
1.22	Names of Fractions	19
1.27	Vocabulary Review	22
1.28	Practice Test	24

Chapter 2 Measurement

2.1	Measuring with a Ruler	29
2.7	Large and Small Measurements	34
2.11	Understanding Area	38
2.15	Understanding Volume	41
2.19	Orientation and Direction	44
2.23	Temperature	47
2.27	Measuring Time	50
2.31	Measuring Weight	53
2.35	Measuring Capacity	55
2.39	Vocabulary Review	58
2.41	Practice Test	60

Chapter 3 Geometric Figures

3.1	Geometric Symbols	65
3.4	Measuring Angles	67
3.8	Identifying Triangles	71
3.12	Quadrilaterals	75
3.16	Drawing Figures with a Protractor and Ruler	79
3.20	Drawing Solid Figures	82
3.25	Classifying Solid Figures	84
3.28	Vocabulary Review	87
3.29	Practice Test	88

Chapter 4 Addition and Subtraction

4.1	Addition and Subtraction	93
4.5	Addition Story Problems	96
4.9	Adding Money	99
4.14	Subtraction Story Problems	102
4.20	Getting Change Problems	105
4.26	Mixed Story Problems	108
4.28	Vocabulary Review	110
4.29	Practice Test	111

Chapter 5 Rounding and Estimating

5.1	Rounding Tens	117
5.6	Rounding Big Numbers	119
5.11	Estimating	123
5.16	Rounding Decimal Fractions	124
5.21	Rounding in Money Problems	127
5.26	Vocabulary Review	129
5.27	Practice Test	130

Chapter 6 Multiplication

6.1	Multiplication	135
6.6	Multiplication Story Problems	138
6.10	Units Story Problems	142
6.14	Multiplying Decimal Fractions	145
6.18	Multiplication Story Problems with Money	149
6.22	Multiplication in Percent Problems	152
6.26	Using Formulas	155
6.30	More Story Problems	157
6.32	Vocabulary Review	158
6.33	Practice Test	159

Chapter 1

Names of Numbers

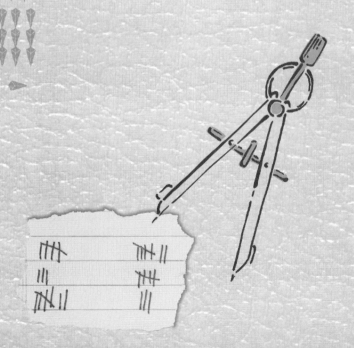
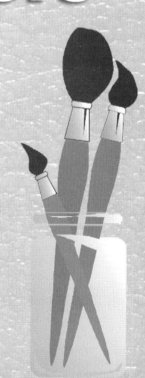

1.1 Names of Numbers

Numbers have names. The names are easy to learn. We use numbers to count things. You will use numbers every day.

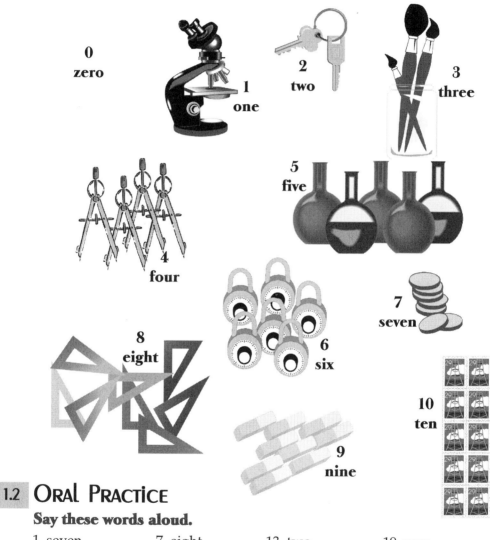

1.2 Oral Practice

Say these words aloud.

1. seven
2. nine
3. four
4. three
5. two
6. five
7. eight
8. ten
9. ten
10. zero
11. one
12. six
13. two
14. five
15. eight
16. six
17. three
18. nine
19. zero
20. four
21. seven
22. ten
23. four
24. five

Names of Numbers

What are these numbers?

5   2   4   9   1   0

3   6   10   8   7

1.3 HOMEWORK

Set A: Spell out the name for the numeral.

1. 3
2. 4
3. 5
4. 4
5. 7
6. 8
7. 6
8. 0
9. 1
10. 10
11. 7
12. 9
13. 3
14. 2
15. 1
16. 5
17. 6
18. 8
19. 9
20. 10
21. 4
22. 0
23. 3
24. 5

Set B: Write the numeral.

25. seven
26. one
27. five
28. two
29. nine
30. zero
31. ten
32. three
33. six
34. four
35. eight
36. one
37. zero
38. one
39. two
40. seven
41. five
42. nine
43. ten
44. six
45. eight
46. one
47. two
48. zero

Set C: Spell out the name for the numeral.

49. 3
50. 1
51. 4
52. 7
53. 2
54. 9
55. 5
56. 8
57. 0
58. 6
59. 10
60. 5
61. 1
62. 0
63. 7
64. 10
65. 9
66. 3
67. 8
68. 4
69. 6
70. 2
71. 5
72. 8
73. 4
74. 7
75. 10
76. 3
77. 1
78. 0
79. 6
80. 2
81. 9
82. 3
83. 0
84. 9

Set D: Write the numeral.

85. ten	103. three
86. seven	104. ten
87. four	105. seven
88. eight	106. zero
89. five	107. one
90. two	108. five
91. six	109. ten
92. four	110. six
93. eight	111. nine
94. zero	112. zero
95. eight	113. three
96. five	114. nine
97. nine	115. two
98. two	116. six
99. seven	117. zero
100. four	118. one
101. one	119. three
102. three	120. nine

1.4 Vocabulary

Count
Homework
Number
Name
Oral
Practice
Problem
Say
Set
Write

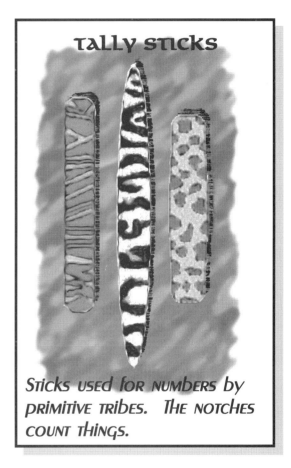

Tally sticks

Sticks used for numbers by primitive tribes. The notches count things.

Names of Numbers

1.5 More Names of Numbers

Large numbers have names. Small numbers have names. You put the names together.

10 ten	25 twenty-five
11 eleven	26 twenty-six
12 twelve	27 twenty-seven
13 thirteen	28 twenty-eight
14 fourteen	29 twenty-nine
15 fifteen	30 thirty
16 sixteen	31 thirty-one
17 seventeen	32 thirty-two
18 eighteen	33 thirty-three
19 nineteen	34 thirty-four
20 twenty	35 thirty-five
21 twenty-one	36 thirty-six
22 twenty-two	37 thirty-seven
23 twenty-three	38 thirty-eight
24 twenty-four	39 thirty-nine

Some numbers have special names (10 to 19). Other numbers put the names together (20 to 99).

40	forty
50	fifty
60	sixty
70	seventy
80	eighty
90	ninety

Big numbers look like this.

49 is forty-nine.

51 is fifty-one.   75 is seventy-five.

67 is sixty-seven.   84 is eighty-four.

99 is ninety-nine.

Names of Numbers

1.6 ORAL PRACTICE

Say these words aloud.
1. twelve
2. sixty-three
3. ninety-six
4. sixty-one
5. seventy-eight
6. twenty-nine
7. thirty-six
8. seventy-seven
9. thirty-two
10. eighty-three
11. twenty-eight
12. fifty-seven
13. forty-eight
14. fifty-four
15. sixteen
16. eighty-nine
17. fifty-one
18. ninety-seven
19. thirty-eight
20. eighty-three

1.7 HOMEWORK

Set A: Spell out the name for the numeral.

1. 15
2. 37
3. 52
4. 79
5. 91
6. 4
7. 25
8. 48
9. 67
10. 89
11. 99
12. 1
13. 10
14. 28
15. 35
16. 49
17. 51
18. 62
19. 73
20. 0
21. 88
22. 31
23. 83
24. 29

Set B: Write the numeral.

25. eighty-nine
26. sixty-two
27. eleven
28. ninety-eight
29. seventy-six
30. twenty-seven
31. seventy-nine
32. thirty-three
33. eighty-six
34. fifty-two
35. thirty-seven
36. fifty-eight
37. sixty-five
38. thirty-one
39. twenty-eight
40. fifteen
41. ninety-seven
42. seventeen

Set C: Spell out the name for the answer.

43. 13 + 62 =
44. 46 + 29 =
45. 38 + 52 =
46. 79 + 12 =
47. 84 + 9 =
48. 17 + 24 =
49. 56 + 32 =
50. 23 + 47 =
51. 63 + 22 =
52. 51 - 37 =
53. 74 - 21 =
54. 45 - 13 =
55. 75 - 18 =
56. 99 - 45 =
57. 88 - 32 =
58. 23 - 11 =
59. 82 - 49 =
60. 43 - 37 =
61. 19 x 3 =
62. 23 x 3 =
63. 8 x 9 =
64. 6 x 7 =
65. 4 x 12 =
66. 3 x 15 =
67. 21 ÷ 3 =
68. 48 ÷ 3 =
69. 50 ÷ 2 =
70. 93 ÷ 3 =
71. 100 ÷ 5 =
72. 120 ÷ 4 =

Names of Numbers

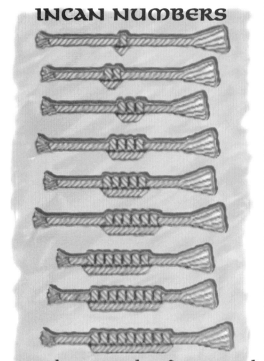

INCAN NUMBERS

The Incas lived in South America. They had no alphabet or numbers. They used strings with knots to count things. A string was called a quipu.

Each town had quipucamayocs, the knotkeepers. They kept the records of the town.

The quipus counted people. They counted taxes. They divided people into groups to work.

The Incas had a big empire. It was very organized even though it had no writing.

Set D: Write the numeral.

73. 85
74. 11
75. 9
76. 79
77. 20
78. 36
79. 24
80. 67
81. 3
82. 12
83. 58
84. 43
85. 4
86. 45
87. 6
88. 92
89. 10
90. 44
91. 5
92. 75
93. 5
94. 86
95. 16
96. 77
97. 1
98. 89
99. 96
100. 24
101. 35
102. 89
103. 10
104. 20
105. 83
106. 74
107. 36
108. 15
109. 27
110. 0
111. 91
112. 30
113. 41
114. 82
115. 12
116. 14
117. 73
118. 5
119. 19
120. 74
121. 9
122. 85
123. 36
124. 77

1.8 Vocabulary

Answer
Big
Large
Look
Say
Small
Special
Together

Names of Numbers

1.9 Even More Names of Numbers

Very large numbers have names. Use the names of small numbers and large numbers.

100	one hundred
200	two hundred
300	three hundred
400	four hundred
500	five hundred
600	six hundred
700	seven hundred
800	eight hundred
900	nine hundred
101	one hundred one
102	one hundred two
110	one hundred ten
111	one hundred eleven
120	one hundred twenty
121	one hundred twenty-one
142	one hundred forty-two

All numbers have names. Even the largest numbers have names.

1,000	one thousand
1,100	one thousand one hundred
1,143	one thousand one hundred forty-three
2,000	two thousand
3,000	three thousand
10,000	ten thousand
20,000	twenty thousand
23,000	twenty-three thousand
37,405	thirty-seven thousand four hundred five
100,000	one hundred thousand
200,000	two hundred thousand

Millions, billions, and trillions are the biggest numbers that we use.

1,000,000	one million
2,000,000	two million
1,000,000,000	one billion
1,000,000,000,000	one trillion

Names of Numbers

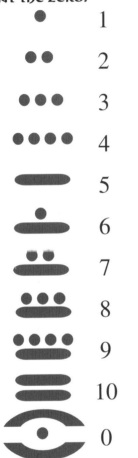

MAYAN NUMBERS

The ancient Mayans of Central America were one of the few cultures to invent the zero.

1.10 Oral Practice

Say these numerals aloud.

1. 673
2. 427
3. 2,785
4. 8,933
5. 5,887
6. 34,174
7. 34,502
8. 16,760
9. 6,691
10. 786
11. 32
12. 5,566
13. 57,763
14. 453
15. 4,891
16. 312,356
17. 450,000
18. 676,905
19. 4,200,000
20. 3,600,500,210
21. 17,323,678,212

Say the answer aloud.

22. 482 + 274 =
23. 379 + 492 =
24. 127 + 510 =
25. 517 + 115 =
26. 850 + 753 =
27. 749 + 891 =
28. 627 + 649 =
29. 245 + 381 =
30. 253 - 59 =
31. 748 - 525 =
32. 612 - 374 =
33. 374 - 127 =
34. 892 - 651 =
35. 581 - 419 =
36. 275 - 116 =

1.11 Vocabulary

All
For
Largest
These
Very
Word
You

Names of Numbers

1.12 HOMEWORK

Set A: Spell out the name for the numeral.
1. 23
2. 657
3. 2,345
4. 7,930
5. 5,899
6. 34,544
7. 12,500
8. 16,980
9. 6,891
10. 599
11. 21,763
12. 5,876
13. 3,333
14. 2,890
15. 112,379
16. 675,985
17. 3,500,000
18. 6,300,575,200

Set B: Write the numeral.
19. two hundred forty-seven
20. three hundred twenty-three
21. two thousand four hundred
22. six thousand eight hundred sixty-seven
23. twelve thousand four hundred thirty-two
24. three hundred seventy thousand
25. four million five hundred sixty thousand
26. twelve million fifty thousand six hundred

Set C: What is the answer? Spell out the name for the numeral.
27. 119 + 733 =
28. 441 + 671 =
29. 708 + 194 =
30. 181 + 282 =
31. 550 + 343 =
32. 802 + 820 =
33. 228 + 451 =
34. 693 + 254 =
35. 937 + 966 =
36. 639 - 569 =
37. 464 - 308 =
38. 740 - 175 =
39. 315 - 102 =
40. 653 - 400 =
41. 376 - 89 =
42. 560 - 137 =
43. 922 - 598 =
44. 874 - 724 =

Set D: Spell out the name for the numeral.
45. 481
46. 527
47. 178
48. 212
49. 664
50. 709
51. 195
52. 353
53. 900
54. 436
55. 8,571
56. 2,497
57. 3,682
58. 2,864
59. 1,053
60. 39,859
61. 52,004
62. 41,961
63. 63,725
64. 74,836
65. 785,711
66. 324,978
67. 236,820
68. 928,645
69. 410,536
70. 1,470,967
71. 3,119,588
72. 6,235,305
73. 2,590,079
74. 4,343,690

Names of Numbers

Set E: Write the numeral.

75. three hundred eighty-one
76. six hundred twenty-six
77. two thousand five hundred thirty-four
78. five thousand one hundred one
79. one thousand six hundred eighty-three
80. forty-nine thousand nine hundred fifty-nine
81. seventy thousand two hundred twenty-one
82. eight hundred seventy thousand one hundred seven
83. nine hundred forty-seven thousand fifty-nine
84. three million five hundred three thousand eight hundred

1.13 Number Periods

The number system is organized into periods. Each period has three digits. A digit is a number with a place value.

Billions Period			Millions Period			Thousands Period			Ones Period		
Hundred Billions	Ten Billions	Billions	Hundred Millions	Ten Millions	Millions	Hundred Thousands	Ten Thousands	Thousands	Hundreds	Tens	Ones

We use commas in numbers. They help us read numbers.

Examples: 1,247

The first comma tells us "thousands".

Names of Numbers

12,498,794

The second comma tells us "millions".

1.14 Oral Practice
What is the place value of the digit "3"?

1. 2,300
2. 12,803
3. 30,002
4. 15,235
5. 13,200
6. 12,573
7. 3,789
8. 29,365
9. 42,736
10. 324,248
11. 632,569
12. 395,470
13. 3,802,468
14. 34,913,574
15. 3,257,456,908

1.15 Homework

Set A: Add commas to these numbers.

1. 7016
2. 3638
3. 6750
4. 14271
5. 902492
6. 51613
7. 857267
8. 108358
9. 399449
10. 2829206
11. 4917184
12. 4005062
13. 53303178
14. 86484959
15. 27565730

Set B: What is the place value of the digit "4"?

16. 60,040
17. 13,304
18. 99,400
19. 124,000
20. 400,000
21. 2,340,000
22. 1,240
23. 4,500
24. 5,400
25. 400,001
26. 456,900
27. 148,987
28. 4,560,333
29. 2,843,987
30. 34,985,003
31. 4,999,000,000
32. 345,125,967,876
33. 411,555,000,000
34. 735,458,200

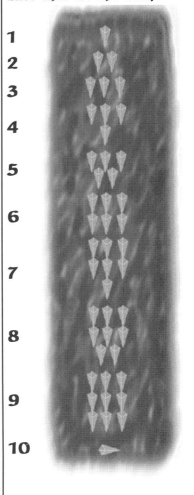

BABYLONIAN NUMBERS

The Babylonians wrote in soft clay with a pointed stick. The marks they made looked like this in the clay.

1
2
3
4
5
6
7
8
9
10

Names of Numbers

Set C: What is the place value of the digit "5"?

35. 5
36. 53
37. 512
38. 5,212
39. 53,980
40. 513,876
41. 5,234,678
42. 53,673,000
43. 4,567
44. 5,678
45. 3,657
46. 2,115
47. 12,567
48. 13,657
49. 15,789
50. 53,467
51. 5,679,000
52. 4,567,098
53. 43,257,098
54. 53,980,000
55. 512,789,342
56. 5,123,876,000
57. 53,986,000,000
58. 534,000,000,000

Set D: What is the place value of the digit "1"?

59. 123
60. 213
61. 321
62. 1,234
63. 2,134
64. 3,214
65. 4,321
66. 12,345
67. 21,345
68. 23,145
69. 23,415
70. 23,451
71. 123,456
72. 213,456
73. 234,561
74. 234,515
75. 1,234,567
76. 2,314,567
77. 2,345,167
78. 12,345,678
79. 23,415,678
80. 213,456,789
81. 921,345,678
82. 1,234,567,890

1.16 Challenge

Change these symbols from the Mayans and Babylonians into numbers.

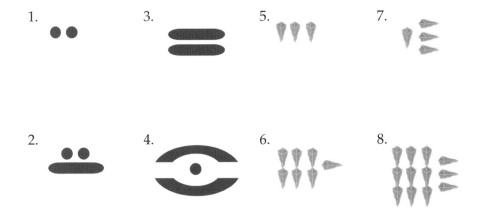

Names of Numbers

Do these problems. Write the answers in the same numerals: Mayan or Babylonian.

9.

12.

10.

13.

11.

14.

1.17 Vocabulary

Clay
Comma
Digit
Example
First
Mark
Number System
Numeral
Organized
Period
Place Value
Second
Soft
Stick
Tell

Names of Numbers

1.18 Names of Decimal Fractions

We use periods in numbers. A period is called a decimal point. It tells us "decimal fractions." Look at this number.

12.3 (twelve and three tenths)

The word **"and"** is read for the decimal point.

Each place in a number has a name.

The names of decimal fractions are like the names of big numbers.

tens 23.4 tenths
twenty-three and four tenths

hundreds 734.83 hundredths
seven hundred thirty-four and eighty-three hundredths

thousands 2,304.014 thousandths
two thousand three hundred four and fourteen thousandths

The names of decimal fractions are easy to remember. You add a "th" to each place name. This is a decimal fraction place name.

0.1	ONE TENTH
0.01	ONE HUNDREDTH
0.001	ONE THOUSANDTH
0.0001	ONE TEN THOUSANDTH
0.00001	ONE HUNDRED THOUSANDTH
0.000001	ONE MILLIONTH

1.19 Oral Practice

Say these words aloud with your teacher.

1. 4.3
2. 17.01
3. 15.39
4. 239.95
5. 125.25
6. 73.43
7. 299.16
8. 988.02
9. 625.13
10. 199.9
11. 25.15
12. 3.125
13. 19.015
14. 16.999
15. 0.238
16. 1.001
17. 129.039
18. 1.111
19. 12.563
20. 1,467.053

Names of Numbers

Homework

Set A: Write the names for these numerals.

1. 15.4
2. 21.7
3. 125.5
4. 99.6
5. 2.2
6. 0.1
7. 39.9
8. 25.3
9. 1,325.8
10. 75,500.5
11. 4.9
12. 6.19
13. 5.13
14. 7.235
15. 9.6
16. 5.35
17. 12.5
18. 45.65
19. 77.77
20. 19.125

Set B: What is the place value of the digit "6"?

21. 63.09
22. 19.614
23. 125.6
24. 986.34
25. 500.46
26. 1,650.013
27. 23.061
28. 165.2
29. 673.34
30. 2,578.056
31. 12.465
32. 26.01
33. 32.62
34. 4.756
35. 53.61
36. 61.99
37. 7.406
38. 863.01
39. 9.642
40. 18.76

Set C: Spell out the name for the numeral.

41. 13.6
42. 53.8
43. 24.2
44. 44.7
45. 35.3
46. 66.52
47. 76.48
48. 21.95
49. 45.06
50. 82.17
51. 971.33
52. 375.48
53. 588.90
54. 782.14
55. 896.29
56. 1,037.503
57. 6,949.648
58. 9,000.759
59. 2,798.162
60. 1,001.271

Greek Numbers

The ancient Greeks used the letters of their alphabet for numbers. The bar on top of the letter meant that it was a number instead of a word.

$\overline{A}$	1	$\overline{I}$	9	$\overline{P}$	80
$\overline{B}$	2	$\overline{K}$	10	$\overline{\Sigma}$	90
$\overline{\Gamma}$	3	$\overline{\Lambda}$	20	$\overline{T}$	100
$\overline{\Delta}$	4	$\overline{M}$	30	$\overline{Y}$	200
$\overline{E}$	5	$\overline{N}$	40	$\overline{\Phi}$	300
$\overline{Z}$	6	$\overline{\Xi}$	50	$\overline{X}$	400
$\overline{H}$	7	$\overline{O}$	60	$\overline{\Psi}$	500
$\overline{\Theta}$	8	$\overline{\Pi}$	70	$\overline{\Omega}$	600

Names of Numbers

Set D: What is the place value of the digit "5"?

61. 10.5
62. 410.5
63. 5.36
64. 2.25
65. 34.5
66. 61.51
67. 826.25
68. 74.735
69. 9.395
70. 508.4
71. 15.081
72. 56.71
73. 262.5
74. 5.83
75. 9.945
76. 35.13
77. 7.652
78. 474.5
79. 88.54
80. 509.4

Set E: Spell out the name for the numeral.

81. 0.9
82. 0.7
83. 0.6
84. 0.5
85. 0.1
86. 0.2
87. 0.31
88. 0.42
89. 8.85
90. 0.35
91. 0.47
92. 0.61
93. 0.89
94. 1.9
95. 5.1
96. 6.2
97. 2.4
98. 4.5
99. 3.3
100. 7.17
101. 8.69
102. 9.22
103. 1.31
104. 2.43
105. 4.58
106. 61.1
107. 74.4
108. 32.6
109. 55.5
110. 83.2
111. 70.15
112. 83.47
113. 284.1
114. 518.3
115. 426.5
116. 295.96
117. 406.18
118. 638.39
119. 984.45
120. 101.07

1.20 PLACE NAME CHART

Copy this chart and fill in the place names.

5,489,368.79015

					hundreds	tens	ones	tenths	hundredths		

Names of Numbers

1.21 Vocabulary

Add
And
Decimal fraction
Decimal point
Easy
Place
Read
Remember
Teacher
With

1.22 Names of Fractions

Fractions are parts of numbers. Each fraction has a name. Many names are easy to make. Some fractions have special names.

Each part of a fraction has a name. The parts are called the numerator and the denominator.

$$\frac{2}{3} \quad \begin{array}{l}\text{numerator} \\ \text{denominator}\end{array}$$

The name of a fraction is two words. First, you say the numerator. It is a number. Then you say the denominator. It is a special name.

fraction	numerator	denominator	name
1/2	one	half	one-half
1/3	one	third	one-third
2/3	two	third	two-thirds
1/4	one	fourth	one-fourth
4/5	four	fifth	four-fifths
5/6	five	sixth	five-sixths
2/7	two	seventh	two-sevenths
7/8	seven	eighth	seven-eighths
5/9	five	ninth	five-ninths
9/10	nine	ten	nine-tenths
7/11	seven	eleventh	seven-elevenths
5/12	five	twelfth	five-twelfths

Names of Numbers

The names of the denominators are easy to make. Most of them have a "th" after the name of the number. Some are special. The special ones are: half, third, fifth, eighth, ninth, and twelfth.

1.23 Oral Practice

Say these words aloud with your teacher.

1. 3/8
2. 1/9
3. 6/8
4. 1/2
5. 2/5
6. 3/7
7. 3/4
8. 5/6
9. 9/10
10. 2/7
11. 2/6
12. 6/10
13. 7/12
14. 1/4
15. 2/3
16. 7/8
17. 9/11
18. 1/3
19. 3/5
20. 6/7

1.24 Homework

Set A: Spell out the name for each of these fractions.

1. 1/2
2. 2/5
3. 3/7
4. 1/3
5. 3/5
6. 6/7
7. 7/9
8. 3/4
9. 5/6
10. 9/10
11. 2/7
12. 3/8
13. 1/9
14. 6/8
15. 2/3
16. 7/8
17. 9/11
18. 2/6
19. 6/10
20. 7/12
21. 1/4
22. 1/5
23. 1/6
24. 5/7

Set B: Spell out the name for each of these fractions.

25. 2/7
26. 3/5
27. 7/10
28. 1/2
29. 2/3
30. 3/8
31. 1/5
32. 5/6
33. 5/9
34. 1/4
35. 7/8
36. 4/5
37. 1/9
38. 4/7
39. 7/9
40. 1/10
41. 9/10
42. 5/7
43. 2/9
44. 3/4
45. 2/5
46. 1/3
47. 5/8
48. 1/6

Roman Numerals

The Romans used letters for numbers.

I = 1
II = 2
III = 3
IV = 4
V = 5
VI = 6
VII = 7
VIII = 8
IX = 9
X = 10
XX = 20
XXX = 30
XL = 40
L = 50
LX = 60
LXX = 70
LXXX = 80
XC = 90
C = 100
D = 500
M = 1,000
$\overline{V}$ = 5,000
$\overline{X}$ = 10,000
$\overline{C}$ = 100,000
$\overline{D}$ = 500,000
$\overline{M}$ = 1,000,000

Can you change these numbers into our numbers?

1. XXIII
2. LIX
3. CXXV
4. CC
5. DCCLXXIV
6. MCCLIX
7. MCMXCII
8. MMD
9. $\overline{D}$DLV
10. CMXCIX

Names of Numbers

49. 3/10	53. 1/8	57. 4/5	61. 1/3
50. 1/7	54. 8/9	58. 7/9	62. 3/10
51. 3/7	55. 1/2	59. 9/10	63. 3/7
52. 6/7	56. 5/6	60. 3/4	64. 6/7

1.25 Vocabulary

Denominator
Fraction
Numerator
Part

1.26 Challenge

Change these into fractions.

1. $\dfrac{\overline{\Delta}}{\overline{I}}$

2. $\dfrac{\overline{Z}}{\overline{\Lambda A}}$

3. $\dfrac{\overline{\Gamma}}{\overline{H}}$

4. $\dfrac{\overline{\Theta}}{\overline{\Pi\Pi}}$

5. $\dfrac{III}{V}$

6. $\dfrac{XXII}{LXXI}$

7. $\dfrac{V}{IX}$

8. $\dfrac{VII}{XII}$

9.

10.

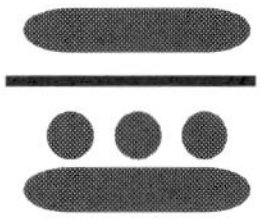

11.

12.

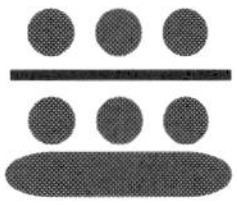

13.

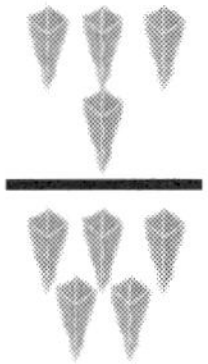

14.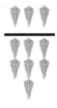

15.

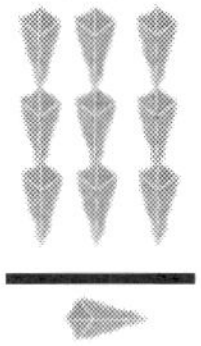

16.

Names of Numbers

1.27 Chapter 1 Vocabulary Review

Use these words to answer the questions and problems below. Look up unfamiliar words in the dictionary. There might be more than one correct answer.

add	fifty	numerator	special
all	first	one	teacher
also	five	oral	tell
and	for	organized	ten
answer	forty	part	tenth
big	four	period	ten thousand
biggest	fourteen	place	these
billion	fourth	place value	third
called	fraction	practice	thirteen
comma	half	problem	thirty
count	homework	read	thousand
decimal	hundred	remember	thousandth
decimal point	hundredth	same	three
denominator	large	say	together
digit	largest	second	trillion
easy	look	seven	twelfth
eight	million	seventeen	twelve
eighteen	name	seventh	twenty
eighth	nine	seventy	two
eighty	nineteen	set	use
eleven	ninth	six	very
eleventh	ninety	sixteen	with
example	number	sixth	word
fifteen	system	sixty	write
fifth	numeral	small	you
			zero

What is the answer? Spell out the name for the numeral using the words above. There might be more than one correct answer.

1. $4 + 3 =$
2. $7 + 2 =$
3. $8 + 2 =$
4. $12 - 4 =$
5. $16 - 2 =$
6. $7 \times 3 =$
7. $2 \times 9 =$
8. $6 \times 8 =$
9. $5 \times 11 =$
10. $100 \div 2 =$
11. $84 \div 4 =$
12. $48 \div 2 =$
13. $112 + 48 =$
14. $56 + 47 =$
15. $936 - 437 =$

Names of Numbers

16. 1,000 = one _____
17. 1,000,000 = ___ _____
18. 43 = _____ - _____
19. 79 = _____ - _____
20. 127 = ___ ___ ___-___
21. 1/2 = one - _____
22. 2/3 = _____ - _____
23. 1/11 = ____ - _____
24. 0.8 = _____ _____
25. 1.2 = _____ and _____
26. 0.25 = _____
27. 14 = _____ _____
28. 40 = _____ _____
29. 414 = ___ _____ _____
30. 440 = ___ _____ _____

31. A _____ helps us read the name of a very big number.

32. The names of _____ numbers go together to make the names of large numbers.

33. In a number, a period is called a _____ _____ .

34. The number 0.87 is called a _____ _____ .

35. The number 2/3 is called a _____.

36. We use _____ to count things.

37. The number system is organized into _____ . (Hint: See page 12)

38. Fractions are _____ of numbers.

39. The parts of a fraction are called the _____ and the _____ .

40. Spell out the name of 2/3 _____ - _____ .

41. Spell out the name of 0.5 _____ _____.

42. Spell out the name of 47 _____ - _____ .

43. Spell out the name of 198 _____ .

44. Spell out the name of 1,275 is _____ .

45. Spell out the name of 2,345,777 _____ .

Names of Numbers

1.28 CHAPTER 1 PRACTICE TEST

Directions: Circle the correct answer.

1. The number 9 is spelled:
   a. eight
   b. nine
   c. four
   d. zero

2. The number 14 is spelled:
   a. forty
   b. ten four
   c. fourteen
   d. one four

3. The number 90 is spelled:
   a. ten-nine
   b. nineteen
   c. ninety
   d. ninety-one

4. The number for sixty is:
   a. 61
   b. 60
   c. 16
   d. 6

5. The number 217 is spelled:
   a. two seventeen
   b. two hundred seventy
   c. twenty one seven
   d. two hundred seventeen

6. The number for four hundred seven is:
   a. 407
   b. 417
   c. 470
   d. 4007

7. The number for twenty-one thousand seven hundred sixty-three is:
   a. 21,763
   b. 217,63
   c. 21763
   d. 2,176.3

8. The number in the hundreds place in 3,748.5901 is:
   a. 0
   b. 7
   c. 4
   d. 9

9. The number in the ten thousands place in 5,789,001 is:
   a. 9
   b. 8
   c. 7
   d. 5

10. Which is the largest?
    a. one million
    b. one billion
    c. one thousand
    d. one trillion

11. Where do the commas go in 84321748?
    a. 84321,748
    b. 843,21,748
    c. 84,321,748
    d. 8,43,21,748

12. What is the place value of the digit "3" in 7,385,700?
    a. hundred thousands
    b. ten thousands
    c. thousands
    d. millions

Names of Numbers

13. The name of the "." in the number 23.17 is:
   a. decimal fraction
   b. decimal point
   c. period
   d. tenths

14. What is the place value of the digit "5" in 3,467.35
   a. hundreds
   b. hundredths
   c. tens
   d. tenths

15. What is the number for two and three hundredths?
   a. 0.23
   b. 2.3
   c. 2.03
   d. 2.003

16. Which number is largest?
   a. 6.75985
   b. 6,759.85
   c. 675.985
   d. 67,598.5

17. The name of 2/3 is:
   a. two-threes
   b. half-threes
   c. two-thirds
   d. three-half

18. The 3 in the number 1/3 is called a:
   a. fraction
   b. numerator
   c. denominator
   d. decimal

19. What is the number for three-sevenths?
   a. 3/7ths
   b. 7/3ths
   c. 7/3
   d. 3/7

20. What is the fraction for IV/IX?
   a. 4/11
   b. 4/9
   c. 5/10
   d. 6/11

Chapter 2

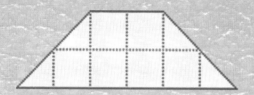

Measurement

Measurement

2.1 Measuring with a Ruler

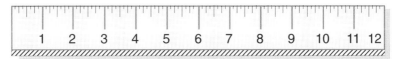

This is a one foot ruler. It is twelve inches long. We use it to measure the length of things. Each inch is divided into fractions. Some of the fractions are: 1/2, 1/4, 3/4, 1/8, 3/8, 5/8, and 7/8.

We use rulers to measure length.
This line is 3 inches long. We also write 3".

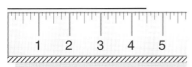

This one is 4 1/2 inches long.

This line is 7/8 of an inch long.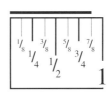

2.2 Practice

Use your ruler to measure these lines.

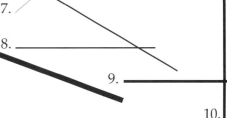

1.
2.
3.
4.
5.
6.
7.
8.
9.
10.

Measurement

A rectangle has four sides.

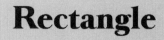

We can measure all four sides.

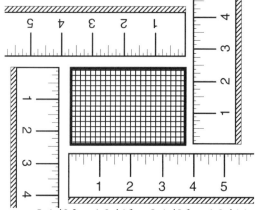

This rectangle measures 2 1/2 by 4 3/4 by 2 1/2 by 4 3/4 inches. Its dimensions are 2 1/2" wide by 4 3/4" long.

2.3 Practice

Find the dimensions of these rectangles.

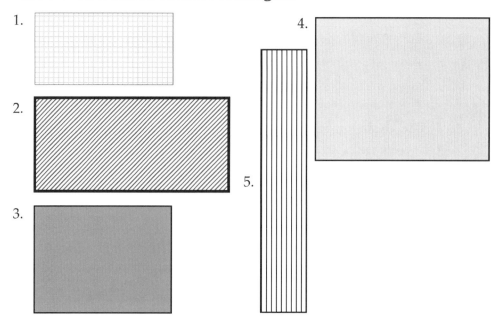

Measurement

A box or a rectangular prism has three dimensions. This one is 6 inches wide by 3 inches high by 8 inches long.

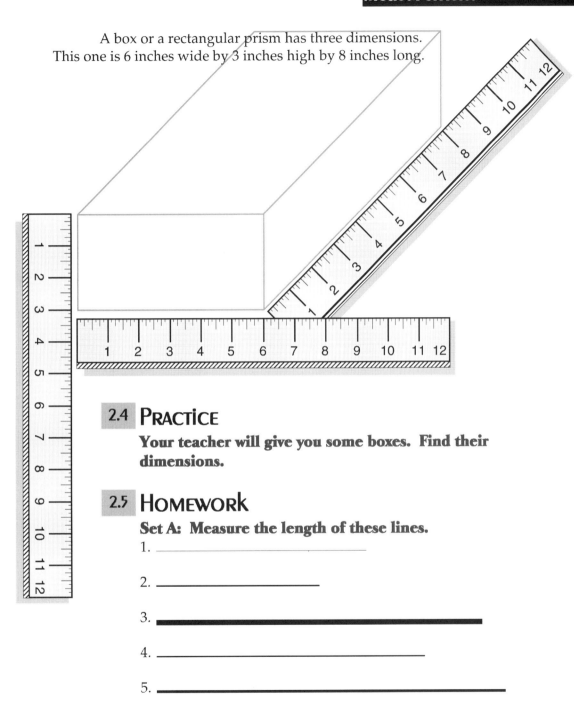

2.4 Practice

Your teacher will give you some boxes. Find their dimensions.

2.5 Homework

Set A: Measure the length of these lines.

1. _____

2. _____

3. _____

4. _____

5. _____

Measurement

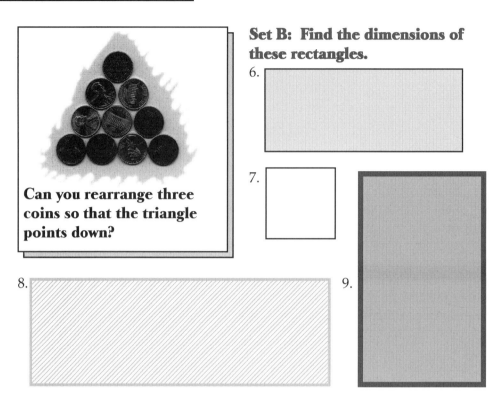

Can you rearrange three coins so that the triangle points down?

Set B: Find the dimensions of these rectangles.

6.

7.

8.

9.

Set C: Find the dimensions of these figures.

10. Equilateral Triangle

11. Pentagon

12. Parallelogram

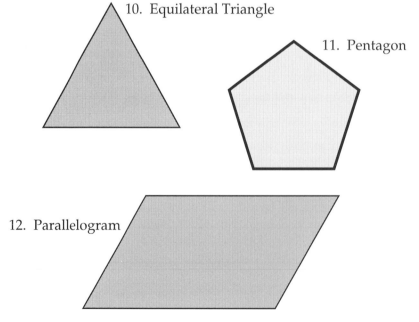

32

Measurement

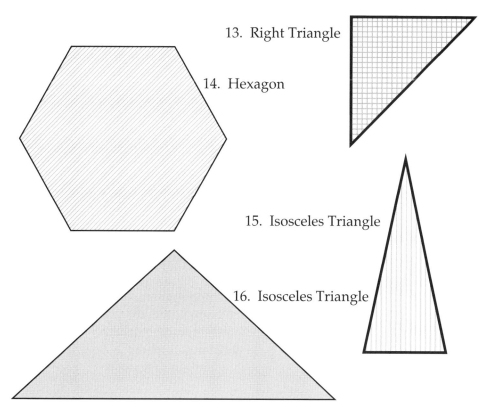

13. Right Triangle
14. Hexagon
15. Isosceles Triangle
16. Isosceles Triangle

2.6 Vocabulary

Box
Dimension
Figure
Foot
High
Inch
Length
Line
Long
Measure
Prism
Rectangle
Ruler
Side
Wide

Measurement

ANCIENT SYMBOLS FOR OPERATIONS

This was used for addition during the Renaissance.

The ancient Greeks used this for subtraction.

This symbol was used by Germans for multiplication during the 1600's.

Division was indicated by this symbol during the 1700's in France.

2.7 Large and Small Measurements

We use large and small units to measure things. These are the units:

12 inches (12 in. or 12")	=	1 foot (1 ft. or 1')
3 feet (3 ft. or 3')	=	1 yard (1 yd.)
1,760 yards (1,760 yd.)	=	1 mile (1 mi.)

You must use an appropriate unit to measure length or distance.

To measure your pencil you can use inches.

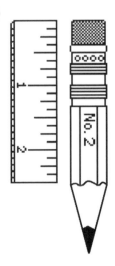

Measurement

To measure your house you can use feet.

To measure a football field you can use yards.

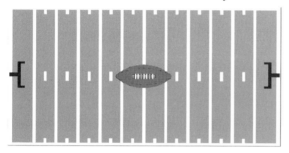

To measure distance to another city you can use miles.

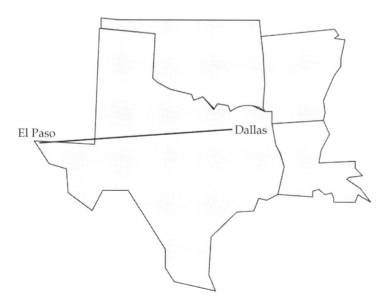

Measurement

2.8 PRACTICE

What unit do you use to measure these things?

1. Your height
2. Your car
3. The distance to school
4. A shoe box
5. A cat
6. A bug or insect
7. Your arm
8. Your waist
9. A candy bar
10. A bicycle
11. A book
12. A pencil
13. A staple
14. A window
15. A desk
16. A notebook
17. A hamburger
18. A pizza
19. A stove
20. A school
21. A classroom
22. United States
23. A sandwich
24. Your finger
25. A can of soda
26. An egg
27. A blackboard
28. A door
29. A piece of gum
30. A plate
31. A glass
32. A spoon

Numbers from Other Languages

English	German	Spanish
zero	null	cero
one	eins	uno
two	zwei	dos
three	drei	tres
four	vier	cuatro
five	fünf	cinco
six	sechs	seis
seven	sieben	siete
eight	acht	ocho
nine	neun	nueve
ten	zehn	diez

Italian	Latin
zero	
uno	unus
due	duo
tre	tres
quattro	quatuor
cinque	quinque
sei	sex
sette	septem
otto	octo
nove	novem
dieci	decem

Measurement

2.9 Challenge

Write the answer in the language of the problem.

Example: due + quattro = sei (from Italian)

1. tres + quatuor =

2. fünf + eins =

3. nove + uno =

4. nueve - cinco =

5. septem - duo =

6. sei - tre =

7. acht - zwei =

8. diez - cuatro =

9. due x tre =

10. uno x cinco =

11. duo x quinque =

12. zwei x fünf =

13. nove ÷ tre =

14. octo ÷ quatuor =

15. diez ÷ dos =

16. sechs ÷ drei =

2.10 Vocabulary

Appropriate
City
Distance
Far
Foot
Feet
Field
Football
House
Inch
Inches
Large
Mile
Pencil
Small
Unit
Yard
Yards

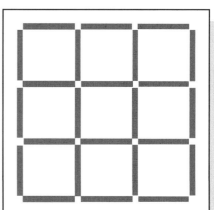

Remove eight sticks and leave just two squares that do not touch.

Measurement

2.11 Understanding Area

Area is the number of squares inside a figure. This rectangle has 28 squares. The area of this rectangle is 28.

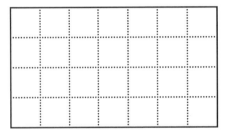

2.12 Practice

What is the area of these figures?

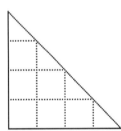

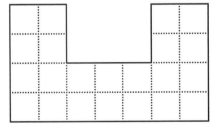

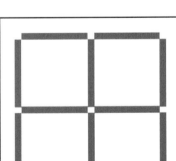

Rearrange two of these objects to make seven squares.

Measurement

The squares can be big or small. The size of the squares is called the unit.

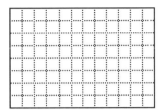

2.13 HOMEWORK

Set A: What is the area of these figures?

1.

2.

3.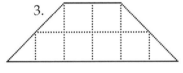

4.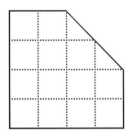

5.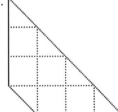

6.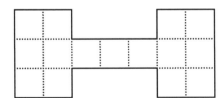

Measurement

Set B: Copy these figures on your paper using one centimeter squares. Draw the squares. What is the area?

7.

8.   9.

Set C: Use 1/4 inch squares.

10.

11.

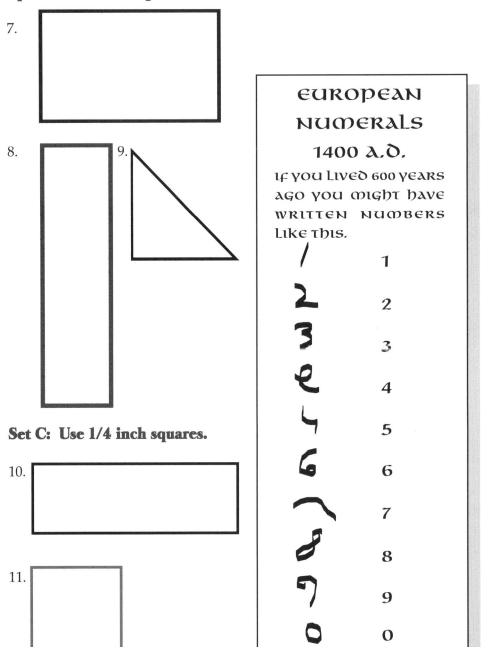

EUROPEAN NUMERALS 1400 A.D.

If you lived 600 years ago you might have written numbers like this.

/	1
ʔ	2
ʒ	3
ε	4
५	5
६	6
）	7
४	8
९	9
o	0

2.14 Vocabulary

Area
Big
Centimeter
Copy
Draw
Figure
Rectangle
Size
Small
Square
Unit square

2.15 Understanding Volume

Volume is how many equal cubes are in a figure. This is a one inch cube. Its volume is one cubic inch.

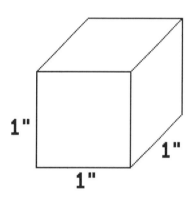

This figure contains two cubes. Its volume is two cubic inches.

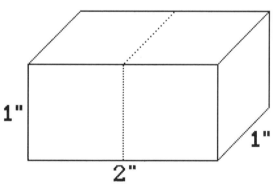

Measurement

2.16 Practice

How many cubes in these figures? What is their volume.

1.
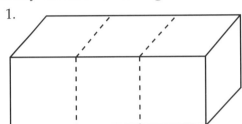

2.
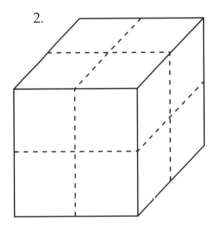

3.
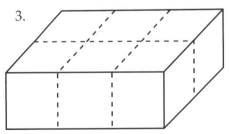

EAST ARABIC NUMERALS 1575 A.D.	
١	1
٢	2
٣	3
٤	4
٥	5
٦	6
٧	7
٨	8
٩	9
٠	0

2.17 Vocabulary

Copy   Leave
Count   Remove
Cube   Square
Cubic inch   Sticks
Draw   Touch
Equal   Volume

Measurement

2.18 Homework

What is the volume? Copy problems 2, 3, and 4, and draw in the squares. Then count them.

1.

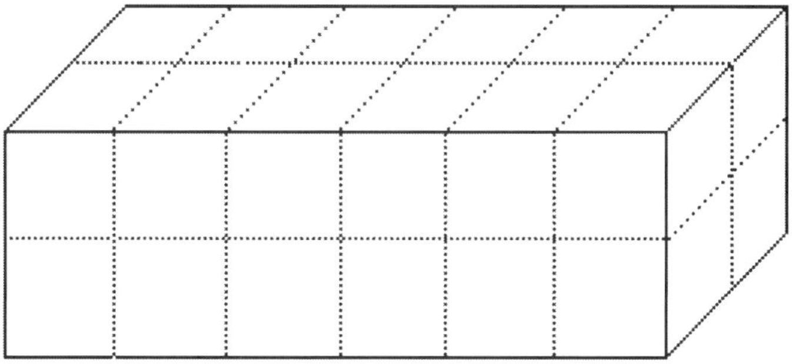

2.

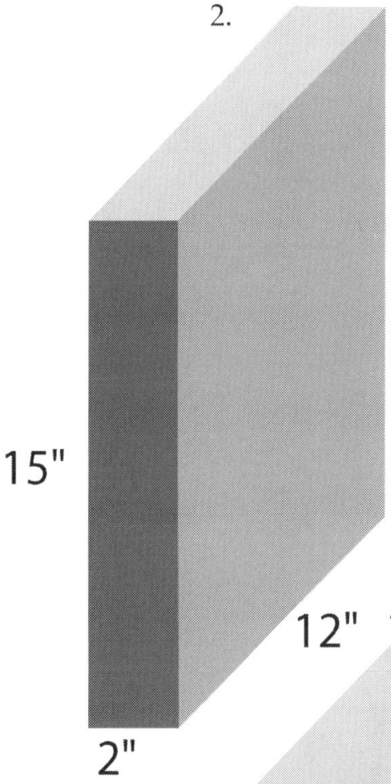

15"

2"

12"

3.

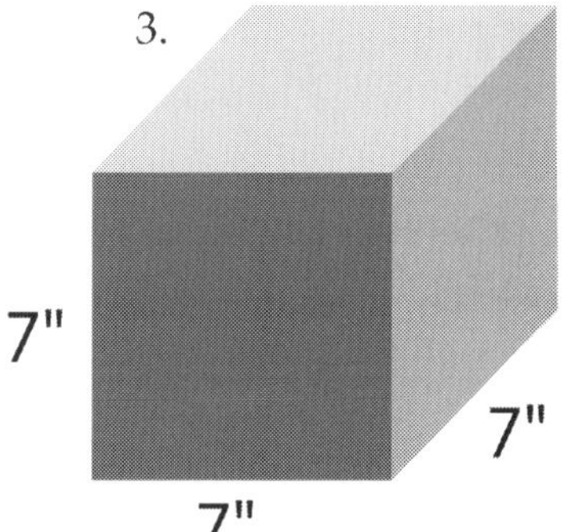

7"

7"

7"

4.

8"

25"

30"

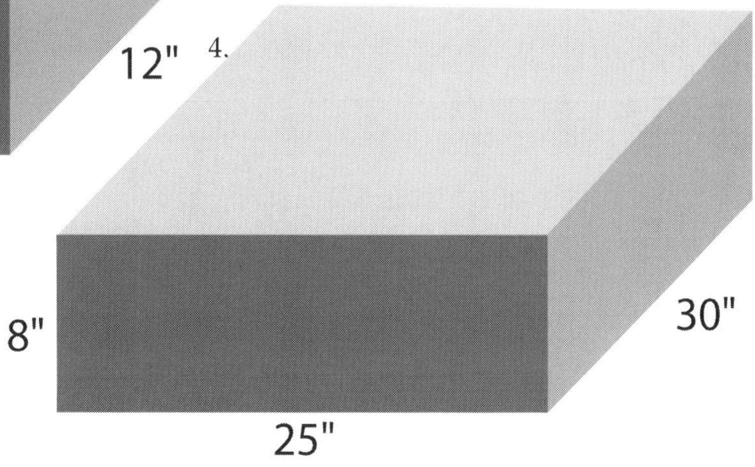

Measurement

Optical Illusions

Some things are not as they seem. Optical illusions make you see things incorrectly.

Which line is longer, A or B?

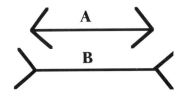

Are the two horizontal lines parallel?

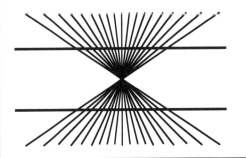

2.19 Orientation and Direction

This is a magnetic compass. It measures direction. The directions are north, south, east, and west.

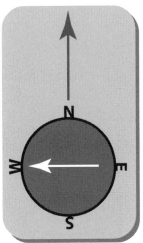

The circle is divided into 360 degrees (360°).

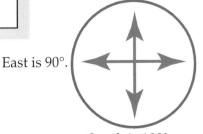

East is 90°. North is 0° West is 270°.

South is 180°.

Then to North again which is 360°.

Measurement

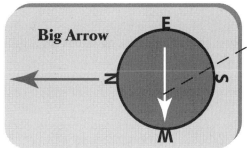

Big Arrow · **Small Arrow**

The important parts of a compass

Dial

You can find the direction of something.

First, point the big arrow at it.

Second, turn the dial to match the little arrow to north (N).

Third, read the direction on the dial.

Read direction here.

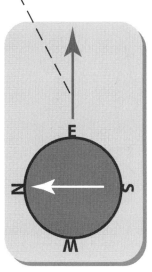

```
┌─────────────────────────────────┐
│      Learning the Months        │
│                                 │
│  The months in English are:     │
│       January                   │
│       February                  │
│       March                     │
│       April                     │
│       May                       │
│       June                      │
│       July                      │
│       August                    │
│       September                 │
│       October                   │
│       November                  │
│       December                  │
│                                 │
│  You can remember the number of │
│  days in each month with a simple │
│  poem:                          │
│                                 │
│    Thirty days hath September,  │
│    April, June, and November.   │
│  February has twenty-eight alone,│
│     All the rest have thrity-one,│
│  Except in Leap Year, that's the time│
│        when February's days     │
│          are twenty-nine.       │
└─────────────────────────────────┘
```

Measurement

2.20 PRACTICE

Use a magnetic compass to measure the direction of these things in your classroom. Your teacher will help.

1. The door
2. The teacher's desk
3. A window
4. The blackboard
5. The file cabinet
6. The bookcase
7. The bulletin board
8. Your best friend
9. The closet
10. The computer
11. The overhead projector
12. The projection screen

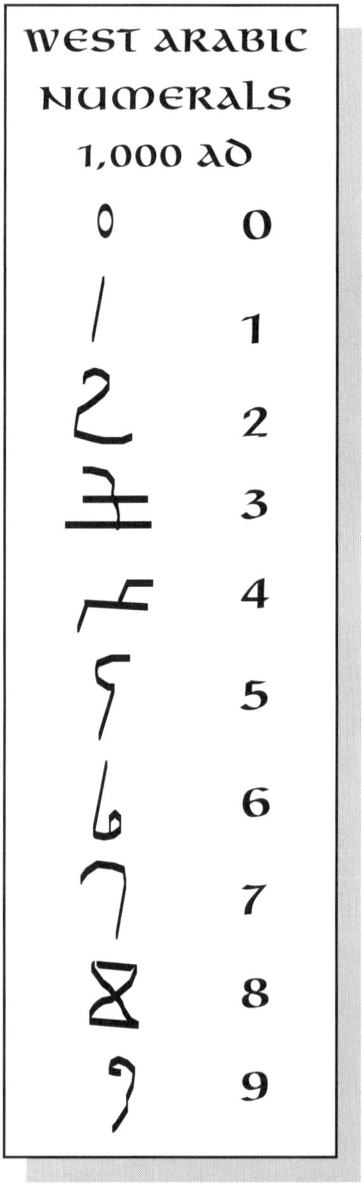

2.21 VOCABULARY

Arrow
Circle
Compass
Degree
Dial
Divide
Direction
East
Magnetic
Match
Measure
North
Point
Read
South
West

WEST ARABIC NUMERALS 1,000 AD

0	0
1	1
2	2
3	3
4	4
5	5
6	6
7	7
8	8
9	9

Measurement

2.22 Challenge

Use the European Numerals from 1400 A.D. (on page 40) and the Ancient Symbols for Operations (on page 34) to solve these problems.

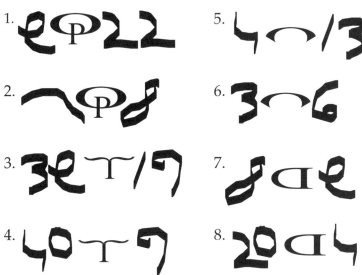

2.23 Temperature

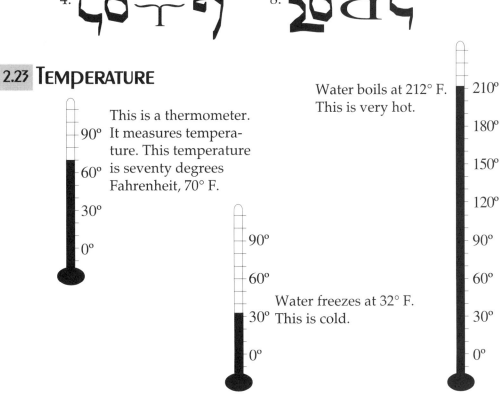

Measurement

More Optical Illusions

Which is longer, line segment AB or CD?

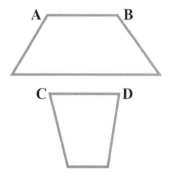

Do these two figures look the same?

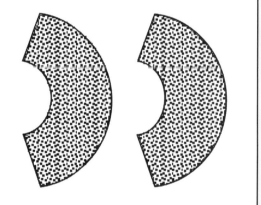

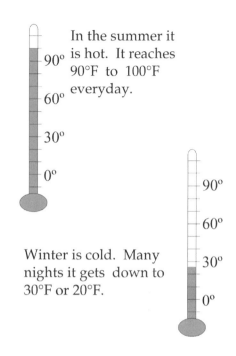

In the summer it is hot. It reaches 90°F to 100°F everyday.

Winter is cold. Many nights it gets down to 30°F or 20°F.

2.21 Practice

Read the temperature on these thermometers. Use the degree sign (°).

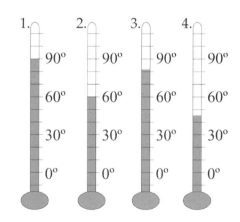

Measurement

Everyday we have a high temperature. We also have a low temperature.

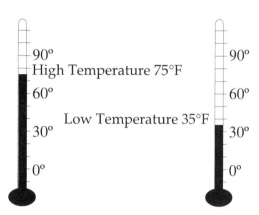

The range of temperatures is the low temperature subtracted from the high temperature.

75°
-35°

40° Range

2.25 Practice

Find the range of high temperatures and low temperatures. Use the degree sign (°) in your answers.

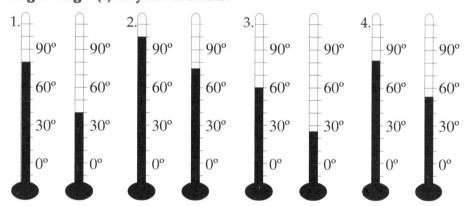

2.26 Vocabulary

Answer	Fahrenheit	Range	Temperature
Boil	Freeze	Reach	Thermometer
Cold	Get	Read	Use
Degree	High	Sign	Very
Down	Hot	Subtract	Water
Everyday	Low	Summer	Winter

Measurement

2.27 Measuring Time

This is a clock. It measures hours, minutes, and seconds. A clock that we wear on the wrist is called a watch.

Clock

Some clocks and watches use hands to tell time. Some use numbers. Most watches and clocks only go to twelve o'clock. Others are twenty-four hour clocks. These are used by the military and airlines.

Hourglass

Grandfather Clock

Cuckoo Clock

This is a calendar. It measures days, weeks, months, and years.

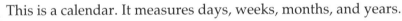

January						
1	2	3	4	5	6	7
8	9	10	11	12	13	14
15	16	17	18	19	20	21
22	23	24	25	26	27	28
29	30	31				

Measurement

You need to know these facts to measure time:

Wristwatch

60 seconds	=	1 minute
60 minutes	=	1 hour
24 hours	=	1 day
7 days	=	1 week
28 to 31 days	=	1 month
365 or 366 days	=	1 year
12 months	=	1 year
52 weeks	=	1 year

Digital Stop Watch

2.28 Practice

Set A: How many seconds in these problems?
1. 7 minutes = ____ seconds
2. 12 minutes = ____ seconds
3. 5 minutes 15 seconds = ____ seconds
4. 13 minutes 45 seconds = ____ seconds

Set B: How many minutes in these problems?
5. 5 hours = ____ minutes
6. 19 hours = ____ minutes
7. 7 hours 20 minutes = ____ minutes
8. 14 hours 35 minutes = ____ minutes

Set C: How many weeks in these problems? Hint: Divide by seven.
9. 21 days = ____ weeks
10. 36 days = ____ weeks ____ days

2.29 Homework

Set A: How many seconds in these problems?
1. 17 minutes
2. 3 minutes
3. 15 minutes
4. 6 minutes 15 seconds
5. 21 minutes 40 seconds
6. 35 minutes 12 seconds
7. 49 minutes 18 seconds

Measurement

Set B: How many minutes in these problems?

8. 5 hours
9. 12 hours
10. 23 hours
11. 24 hours
12. 1 day
13. 2 days
14. 6 hours 15 minutes
15. 2 days 3 hours

Set C: How many weeks in these problems?

16. 14 days
17. 91 days
18. 65 days
19. 1 month
20. 3 months
21. 3 years

Set D: How many days in these problems?

22. 48 hours
23. 168 hours
24. 125 hours
25. 3 weeks
26. 7 weeks
27. 12 weeks
28. 2 months
29. 6 months
30. 2 years
31. January
32. February
33. September
34. March and April
35. June, July, and August
36. September to December
37. 2 years 3 weeks & 6 days
38. 3 years 4 weeks & 3 days
39. 5 years 6 weeks and 1 day
40. 10 years 5 weeks and 4 days
41. 21 years 20 weeks and 2 days
42. 25 years 10 weeks and 5 days

2.30 Vocabulary

Airline
Calendar
Clock
Day
Divide
Hand
Hour
Military
Minute
Month
Multiply
O'clock
Second
Time
Watch
Wear
Week
Wrist
Year

Measurement

2.31 Measuring Weight

The basic units of weight are ounces (oz.), pounds (lb.), and tons.

$$16 \text{ oz.} = 1 \text{ lb.}$$
$$2{,}000 \text{ lb.} = 1 \text{ ton}$$

We use ounces to measure small amounts. We measure food in small packages and cans with ounces.

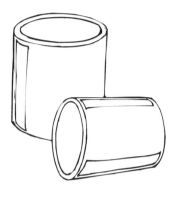

We use pounds to weigh heavier things. We use pounds to weigh people, animals, meat, fruit, and vegetables.

Some things weigh more than one pound. To change from ounces to pounds, you have to divide. The **R** in these problems stands for a remainder in the division problem.

$$20 \text{ oz.} = 1 \text{ lb. } 4 \text{ oz.}$$
$$(20 \div 16 = 1 \text{ R } 4)$$

$$16 \overline{)20} \underline{1 \text{ Remainder } 4}$$

$$60 \text{ oz.} = 3 \text{ lbs. } 12 \text{ oz.}$$
$$(60 \div 16 = 3 \text{ R } 12)$$

We use tons to weigh very large amounts of materials. We weigh very heavy objects with tons. We use tons to weigh sand, dirt, steel, rocks, grain, trucks, and ships.

Measurement

2.32 Practice

Set A: Change to ounces. Hint: Multiply by sixteen.
1. 2 lbs.
2. 5 lbs., 6 oz.
3. 16 lbs.
4. 13 lbs., 12 oz.

Set B: Change to pounds.
5. 2 tons
6. 5 tons
7. 9 tons
8. 3 tons, 500 lbs.
9. 7 tons, 1,225 lbs.
10. 80 oz.
11. 48 oz.
12. 144 oz.
13. 108 oz.
14. 134 oz.

2.33 Homework

Set A: Change to ounces.

1. 3 lbs.
2. 9 lbs.
3. 21 lbs.
4. 5 lbs.
5. 15 lbs.
6. 100 lbs.
7. 2 lbs., 6 oz.
8. 17 lbs., 11 oz.
9. 1 lb., 2 oz.
10. 3 lbs., 5 oz.
11. 10 lbs., 10 oz.
12. 21 lbs., 15 oz.

Set B: Change to pounds.

13. 3 tons
14. 7 tons
15. 11 tons
16. 112 oz.
17. 96 oz.
18. 64 oz.
19. 2 tons, 600 lbs.
20. 5 tons, 1,325 lbs.
21. 7 tons, 250 lbs.
22. 151 oz.
23. 38 oz.
24. 75 oz.

Set C: Add, subtract, or multiply, and then change your answer to pounds.

25. 14 oz. + 12 oz. =
26. 24 oz. + 8 oz. =
27. 35 oz. - 19 oz. =
28. 83 oz. - 37 oz. =
29. 8 oz. x 6 cans =
30. 24 oz. x 8 cans =
31. 3 lbs., 7 oz. + 9 lbs. ,9 oz. =
32. 16 lbs., 13 oz. + 15 lbs., 12 oz. =
33. 93 lbs., 15 oz. + 14 lbs., 11 oz. =
34. 5 lbs., 12 oz. - 3 lbs., 8 oz. =
35. 19 lbs., 14 oz. - 7 lbs., 14 oz. =
36. 21 lbs., 12 oz. - 9 lbs., 13 oz. =
37. 16 lbs., 10 oz. - 9 lbs., 15 oz. =
38. 4 lbs., 3 oz. x 3 =
39. 5 lbs., 8 oz. x 4 =
40. 3 tons, 250 lbs. x 8 =

Measurement

2.34 Vocabulary

Amount	Pound
Animal	Ounce
Can	Remainder
Dirt	Rock
Food	Sand
Fruit	Ship
Grain	Small
Heavier	Steel
Heavy	Thing
Large	Ton
Materials	Truck
Meat	Vegetable
More than	Weigh
Package	Weight
People	

2.35 Measuring Capacity

Pints (pt.), quarts (qt.), and gallons (gal.) are measures of capacity. Capacity measures are for liquids like milk, water, coffee, and tea.

You need to remember these facts:

```
2 pints   = 1 quart
2 quarts  = 1 half-gallon
4 quarts  = 1 gallon
8 pints   = 1 gallon
4 pints   = 1 half-gallon
```

Measurement

We use teaspoons (t.), tablespoons (T.), and cups (C.), when we cook. Teaspoons and cups can be divided into fractions. Three teaspoons equal one tablespoon.

2.36 Practice

Set A: How many pints in each problem?
1. 3 quarts
2. 2 gallons
3. 3 half-gallons
4. 5 gallons
5. 1 gallon, 2 quarts
6. 2 gallons, 1 half-gallon
7. 5 gallons, 1 pint
8. 5 gallons, 1 half-gallon, 7 quarts

Set B: How many gallons in each problem?
9. 32 pints
10. 16 quarts
11. 7 half-gallons
12. 35 pints
13. 4 quarts, 8 pints
14. 5 half-gallons, 2 quarts
15. 3 half-gallons, 3 quarts
16. 3 half-gallons, 6 quarts, 14 pints

Measurement

2.37 HOMEWORK

Set A: How many pints in each problem?

1. 2 qt.
2. 3 gal.
3. 6 half-gallons
4. 2 gal., 3 qt.
5. 3 gallons, 2 quarts
6. 1 half-gallon, 1 quart
7. 6 gallons, 1 half-gallon, 6 quarts
8. 10 gallons, 2 half-gallons, 3 quarts

Set B: How many gallons in each problem?

9. 48 pt.
10. 20 qt.
11. 12 half-gallons
12. 41 pt.
13. 13 qt.
14. 9 half-gallons
15. 173 pt.
16. 29 qt.
17. 25 half-gallons
18. 4 qt., 8 pt.
19. 9 qt., 4 pt.
20. 3 half-gallons, 6 qt.
21. 2 half-gallons, 8 quarts, 24 pints
22. 3 half-gallons, 3 quarts, 6 pints
23. 5 half-gallons, 2 quarts, 8 pints
24. 4 half-gallons, 9 quarts, 3 pints
25. 7 half-gallons, 2 quarts, 4 pints
26. 12 half-gallons, 6 quarts, 12 pints

Set C: How many teaspoons in each problem?

27. 5 T.
28. 3 T.
29. 11 T.
30. 6 1/2 teaspoons
31. 19 1/4 teaspoons

Set D: Add and then change your answer to gallons.

32. 3 quarts + 7 quarts =
33. 2 half-gallons, 3 quarts + 3 half-gallons, 5 quarts =
34. 7 quarts, 9 pints + 8 quarts, 11 pints =

2.38 VOCABULARY

Capacity
Coffee
Cook
Cup
Divide
Equal
Facts
Fractions
Gallon

Half-gallon
Liquid
Milk
Pint
Quart
Tablespoon
Tea
Teaspoon
Water

Measurement

2.39 Chapter 2 Vocabulary Review

Use these words to answer the questions and problems below. Look up unfamiliar words in the dictionary. There might be more than one correct answer.

amount	divide	mile	square
area	down	minute	summer
arrow	draw	month	tablespoon
boil	east	night	teaspoon
box	eight	north	temperature
calendar	equal	o'clock	thermometer
capacity	everyday	orientation	thousand
centimeter	Fahrenheit	ounce	three
circle	figure	pint	time
clock	foot/feet	pound	ton
cold	freeze	pencil	two
compass	gallon	prism	units
cook	half-gallon	quart	volume
cube	hot	range	watch
cubic	hour	rectangle	week
cup	house	rectangular	weigh
day	inch/inches	ruler	weight
degree	length	second	west
dial	line	side	winter
dimensions	long	size	wrist
direction	magnetic	small	yard
distance	measure	south	year

1. A one foot ruler is twelve_____ long.
2. _____ feet equal one yard.
3. A _____ has four sides.
4. _____ are for measuring distance to cities.
5. Area is how many _____ are inside a figure.
6. _____ is how many equal cubes are in a figure.
7. A _____ measures direction.
8. The directions are north, _____, east, and _____.
9. A _____ measures temperature.
10. In the summer it is _____ everyday.
11. A clock measures _____, minutes, and _____.
12. A _____ is a clock on your wrist.
13. A calendar measures _____, weeks, _____, and years.
14. Sixty _____ equals one minute.
15. 365 or 366 days equals one _____.

Measurement

16. Sixteen ounces equals one _____.
17. _____ _____ pounds equals one ton.
18. Pints, _____, and _____ are measures of capacity.
19. _____ pints equals one gallon.
20. We use a _____ to measure length.

2.40 Answers to the Puzzles

From Section 2.5

From Section 2.10

From Section 2.12

Measurement

2.41 Chapter 2 Practice Test
Directions: Circle the correct answer.

1. How many inches are in two feet?
   a. 12
   b. 18
   c. 20
   d. 24

2. How many feet are in five yards?
   a. 3
   b. 15
   c. 9
   d. 8

3. Which is the longest?
   a. a foot
   b. an inch
   c. a mile
   d. a yard

4. Which is East on a magnetic compass?
   a. 90°
   b. 180°
   c. 270°
   d. 360°

5. Which measures direction?
   a. a ruler
   b. a thermometer
   c. a compass
   d. a cup

6. What is the range of temperatures between 35° and 65°?
   a. 100°
   b. 30°
   c. 35°
   d. 90°

7. How many seconds in three minutes?
   a. 3
   b. 60
   c. 180
   d. 240

8. What do we use to measure time?
   a. a watch
   b. a ruler
   c. a compass
   d. a gallon

9. How many days in January?
   a. 28
   b. 29
   c. 30
   d. 31

10. How many ounces in two pounds?
    a. 4
    b. 16
    c. 24
    d. 32

Measurement

11. Which is a measurement of weight?
    a. a ton
    b. a mile
    c. a foot
    d. a degree

12. How many pints in three gallons?
    a. 8
    b. 24
    c. 32
    d. 48

13. Which is a measurement of capacity?
    a. an inch
    b. a quart
    c. a pound
    d. a degree

14. What do we use to measure area?
    a. pints
    b. thermometers
    c. degrees
    d. squares

15. How many ounces are in 2 lbs., 3 oz.?
    a. 5 oz.
    b. 6 oz.
    c. 32 oz.
    d. 35 oz.

16. How many pints are in 2 gallons, 3 quarts?
    a. 5 pints
    b. 6 pints
    c. 16 pints
    d. 22 pints

17. How many feet are in 3 yd., 2 ft.?
    a. 11 ft.
    b. 9 ft.
    c. 6 ft.
    d. 5 ft.

Chapter 3

Geometric Figures

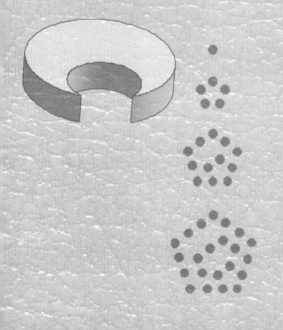
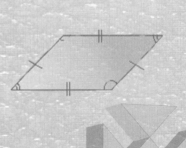

Geometric Figures

3.1 Geometric Symbols

We use many symbols in mathematics. You know some symbols. Some of them follow:

+	1	2.5	%	≠
-	=	1,473	$	1/4
x	>	<	¢	÷

We use symbols in geometry. Geometry is the study of shapes.

This is a line.

⟵―――――――――⟶ A

It is called line A.

These lines are parallel. They do not touch.

We write: P||Q. Line P is parallel to line Q.

These lines cross. The angle is 90°.

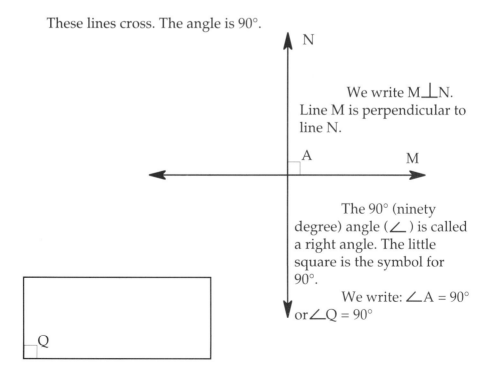

We write M⊥N. Line M is perpendicular to line N.

The 90° (ninety degree) angle (∠) is called a right angle. The little square is the symbol for 90°.

We write: ∠A = 90° or ∠Q = 90°

65

Geometric Figures

Move one toothpick so that the house points east instead of west.

We use symbols to say equal sides. We say: Side a equals side c. We write: a = c.

We say all sides are equal (below).

We say: Side p equals side r and side q equals side s. We write: p = r and q = s.

We use symbols to say equal angles. We say: Angle A equals angle B. We write: ∠A = ∠B.

We say: Angle M equals angle K and angle N equals angle L. We write: ∠M = ∠K and ∠N = ∠L.

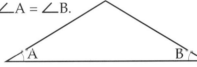

3.2 HOMEWORK

Fill in the blanks. Use words or symbols.

1. 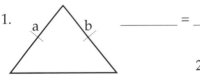 _____ = _____

2. Line _ is _____ to line Q.

3. ∠A = _____
   ∠B = _____

Geometric Figures

4. 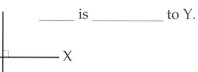 _____ is _____ to Y.

5. Side _____ equals _____ n.

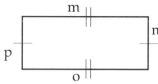

6. ∠P = _____
∠Q = _____
c = _____
Side _____ equals _____ b.

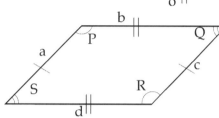

3.3 Vocabulary

Angle
Cross
Degrees
Line
Parallel
Perpendicular
Right
Sides
Symbol

3.4 Measuring Angles

We use a protractor to measure angles. It measures them in degrees (°). Look at the size of these angles.

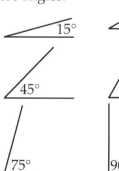

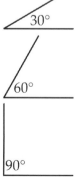

Hindu Numerals
300 B.C.

Symbol	Number
—	1
=	2
≡	3
¥	4
ካ	5
6	6
7	7
5	8
?	9

Geometric Figures

A protractor will help you measure angles. It has two scales. One begins on the right. The other begins on the left.

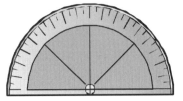

To measure an angle, first put the hole on the point of the angle.

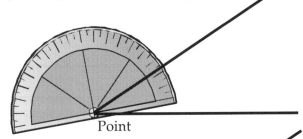

Then line up the baseline with one side of the angle.

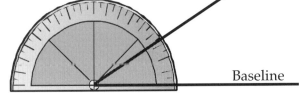

Then measure from the zero on the baseline to the other line of the angle. Read the angle.

Square Numbers

The product when both factors are the same is a square number.

Square numbers are made by making squares with dots. The first four are below. Can you find the next three.

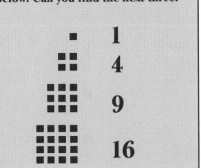

Geometric Figures

3.5 Practice

Set A: Measure these angles with your protractor.

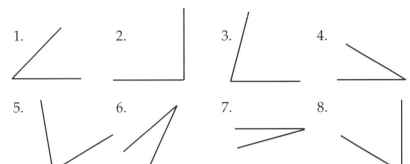

Set B: Draw these angles with your protractor.
9. 35°
10. 49°
11. 65°
12. 95°
13. 85°

3.6 Homework

Set A. Measure these angles with your protractor.

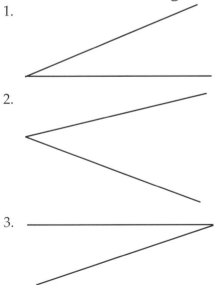

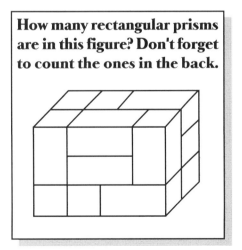

How many rectangular prisms are in this figure? Don't forget to count the ones in the back.

Geometric Figures

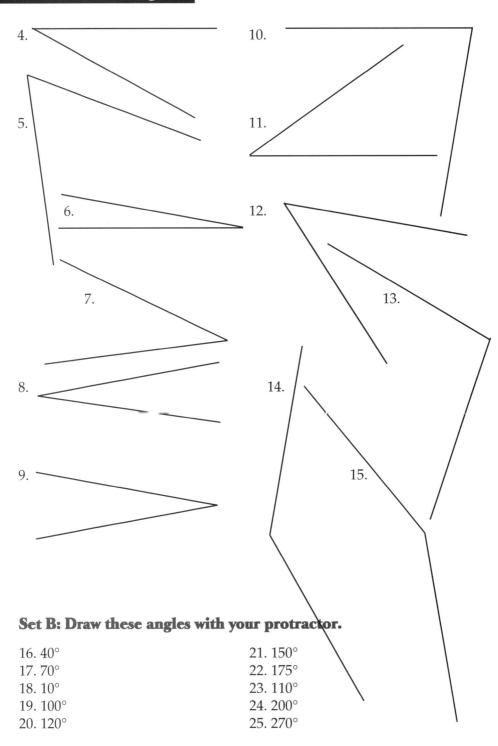

Set B: Draw these angles with your protractor.

16. 40°
17. 70°
18. 10°
19. 100°
20. 120°

21. 150°
22. 175°
23. 110°
24. 200°
25. 270°

3.7 Vocabulary

Angle
Base line
Degrees
Draw
Hole
Left
Line up
Point
Protractor
Right

3.8 Identifying Triangles

We classify triangles into six kinds: right triangles, obtuse triangles, acute triangles, scalene triangles, isosceles triangles, and equilateral triangles. We can remember these better if we use a memory chart like the one below.

R — Right Triangle: One right angle (90°)

O — Obtuse Triangle: One angle more than 90°

A — Acute Triangle: All angle less than 90°

S — Scalene Triangle: No equal sides

I — Isosceles Triangle: Two equal sides

E — Equilateral Triangle: All sides equal

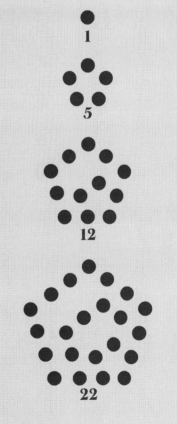

Pentagonal Numbers

Pentagonal Numbers are made by drawing bigger and bigger pentagons with dots. These are the first four.

1
5
12
22

Can you find the next three?

71

Geometric Figures

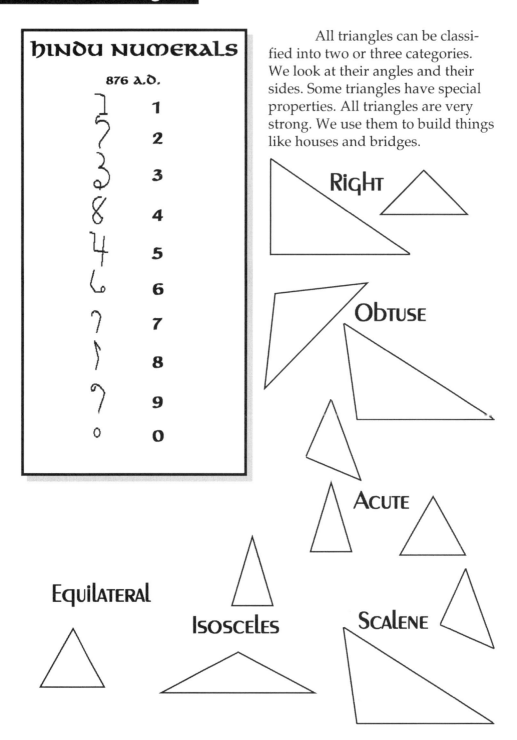

HINDU NUMERALS
876 A.D.

All triangles can be classified into two or three categories. We look at their angles and their sides. Some triangles have special properties. All triangles are very strong. We use them to build things like houses and bridges.

Right

Obtuse

Acute

Equilateral

Isosceles

Scalene

Geometric Figures

3.9 PRACTICE

Make a chart like the one below. Fill in the information from this lesson. It is called a characteristics matrix.

Geometric Figures

Name	Drawing	Sides	Angles
Right Triangle			One 90° angle
		2 equal	2 equal
		No sides equal	Other 2 not equal

How many triangles are in this figure. Be sure to count small ones and large ones.

Geometric Figures

3.10 Homework

Identify these triangles. Use all the names that can apply. You may use a protractor.

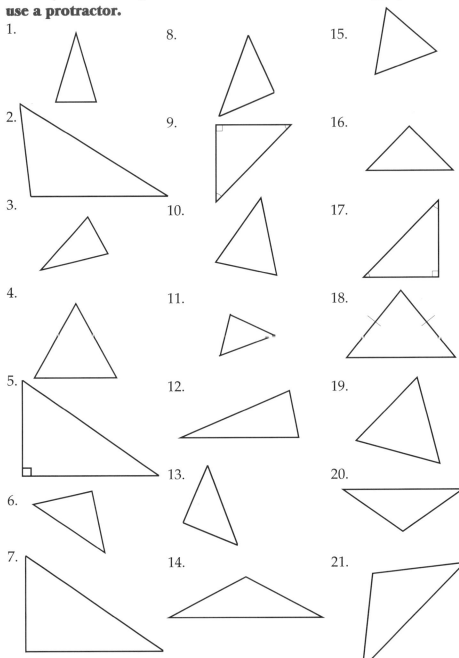

Geometric Figures

3.11 Vocabulary

Acute triangle
Bridges
Build
Category
Characteristic
Classify
Equilateral triangle
Isosceles triangle
Matrix
Memory chart
Obtuse triangle
Right triangle
Scalene triangle
Triangle

Hindu Numerals
1,000 A.D.

૧	1
૨	2
૩	3
૪	4
૫	5
૬	6
૭	7
૮	8
૯	9
०	0

3.12 Quadrilaterals

Quadrilaterals are figures with four sides. There are four kinds: Irregular Quadrilaterals, Trapezoids, Isosceles Trapezoids, Parallelograms, Rhombuses, Rectangles, and Squares.

Irregular Quadrilaterals

Irregular quadrilaterals have four sides. No sides are equal. No angles are equal and no sides are parallel. They can be many different shapes.

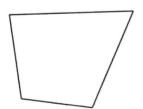
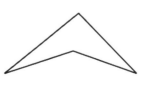
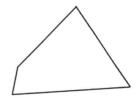

Trapezoids

Trapezoids have only two sides parallel. We call these the two bases. Sides and angles do not have to be equal.

Isosceles Trapezoids

Isosceles trapezoids have two sides equal. They also have some angles equal.

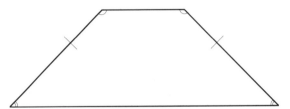

Parallelograms

Parallelograms have opposite sides equal and parallel. They also have opposite angles equal.

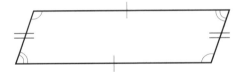

Rhombuses

Rhombuses are parallelograms with all four sides equal.

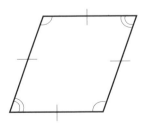

Geometric Figures

Rectangles

Rectangles have opposite sides equal and parallel. All angles are 90°. Rectangles are parallelograms.

Squares

Squares are rectangles with all four sides equal. They are also rhombuses and parallelograms.

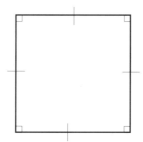

3.13 Practice

Set A: Copy this onto your paper and identify the triangles and quadrilaterals in it.

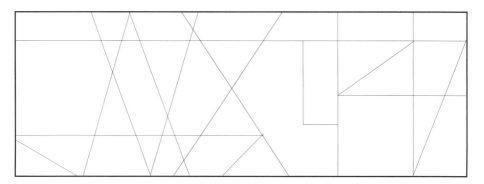

Geometric Figures

Set B: Make a memory chart like the one below. Fill in the information from this lesson. It is called a characteristics matrix.

Geometric Figures

Name	Drawing	Sides	Angles

3.14 Homework

Identify these figures.

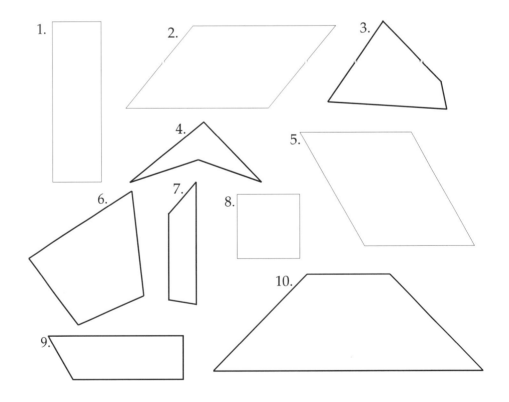

78

Geometric Figures

3.15 Vocabulary

Base
Figure
Irregular quadrilateral
Isosceles trapezoid
Parallelogram
Quadrilateral
Rhombus
Rectangle
Square
Trapezoid

3.16 Drawing Figures with a Protractor and Ruler

The figures that we have studied are easy to draw. You will need a protractor and a ruler. You need to know the facts below. They tell you how to draw regular polygons.

Regular polygons have all angles equal.

Figure	Number of Angles	Size	Total of All Angles
Triangle	3	60°	180°
Square	4	90°	360°
Pentagon	5	108°	540°
Hexagon	6	120°	720°
Octagon	8	135°	1080°

To draw a regular polygon you need to know how long to make each side and the number of degrees in each angle.

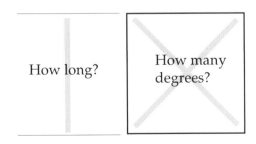

How long?   How many degrees?

Geometric Figures

Let's draw a triangle with each side equal to three inches. First draw a line three inches long.

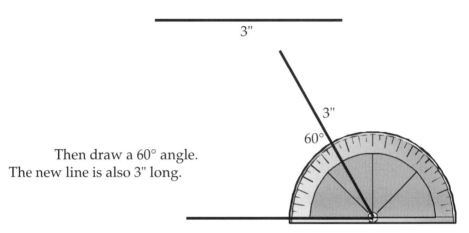

Then draw a 60° angle. The new line is also 3" long.

Then draw the last line. It will also be 3" long if your others are correct.

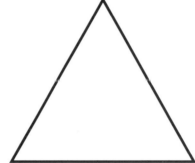

For other shapes you will have to make more angles and sides.

3.17 PRACTICE

Draw a pentagon with 3" sides. Follow these steps.

First, in the middle of your paper, draw a 3" line.

Second, make an angle of 108°. Draw the second side 3" long.

Third, draw another angle of 108°. Draw the third side 3" long.

Fourth, draw the next angle of 108°. Draw the fourth side 3" long.

Last, draw the fifth line to connect the first line and the fourth line.

Geometric Figures

3.18 Homework

Draw one triangle, one square, one pentagon, one hexagon, and one octagon. Make all the sides 2" long.

3.19 Vocabulary

Angle
Degrees
Draw
Figure
Hexagon
Line
Octagon
Pentagon
Polygon
Protractor
Regular
Ruler
Sides
Square
Triangle

Names of Operations

English
add   subtract
multiply   divide

Latin
addĕre   deducĕre
multiplicare dividĕre

Italian
aggiungere   sottarre
moltiplicare dividere

Spanish
sumar   restar
multiplicar dividir

German
addieren   subtrahieren
multiplizieren dividieren

Geometric Figures

3.20 Drawing Solid Figures

You can learn to draw solid figures on paper. You have to learn a few tricks. These will make your figures look real. The first thing is to draw a cube.

To do this first, draw a square.

Then draw three diagonal lines from the corners.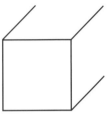
They should all be parallel.

Last, connect the diagonal lines in the back. This is a cube.

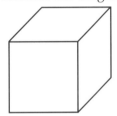

We can make many figures with cubes.

Geometric Figures

3.21 PRACTICE
Draw these shapes.

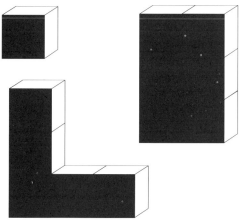

Now draw five shapes using five cubes in each one.

3.22 HOMEWORK
Draw as many shapes as you can, but they all must have only seven cubes.

3.23 CHALLENGE
Use the Names of Operations **(Section 3.17)** and the Numbers from Other Languages **(Section 2.8)** to do these problems. Write the answer in the same language.

Example: acht teilen vier = zwei (German)
1. eins addieren drei
2. acht subtrahieren vier
3. zehn dividieren zwei
4. sex adděre duo
5. novem deducěre tres
6. unus multiplicare quattuor
7. octo dividěre duo
8. sette aggiungere uno
9. otto sottarre tre
10. due moltiplicare cinque
11. dieci dividere due

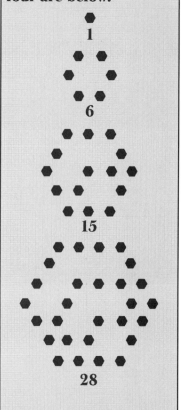

Hexagonal Numbers

Hexagonal numbers are made by drawing bigger and bigger hexagons with dots. The first four are below.

1
6
15
28

Can you find the next three?

Geometric Figures

3.24 Vocabulary

Corner
Cube
Diagonal
Draw
Figure
Real
Shape
Solid
Trick

3.25 Classifying Solid Figures

Solid figures can be classified into five kinds:
- Regular Polyhedrons
- Prisms and Cylinders
- Pyramids and Cones
- Spheres and Ovoids
- Composite Figures

All solid figures have surfaces like triangles, squares, or curved surfaces. Most solid figures have vertices. Vertices are the corners that stick out on a figure.

Regular Polyhedrons

	Name	Vertices	Surfaces
1.	Tetrahedron	4	4 triangles
2.	Cube	8	6 squares
3.	Octahedron	6	8 triangles
4.	Dodecahedron	20	12 pentagons
5.	Icosahedron	18	20 triangles

1.

3.

5.

2.

4.

Geometric Figures

Prisms and Cylinders

	Name	Vertices	Surfaces
6.	Triangular Prisms	6	2 triangles & 3 rectangles
7.	Rectangular Prisms	8	6 rectangles
8.	Pentagonal Prisms	10	2 pentagons & 5 rectangles
9.	Cylinders	0	2 circles & 1 rectangle

6. 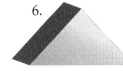 7.  8.   9.

Pyramids and Cones

	Name	Vertices	Surfaces
10.	Triangular Based Pyramids	4	4 triangles
11.	Square Based Pyramids	5	4 triangles & 1 square
12.	Cones	1	1 circle & 1/2 circle

10. 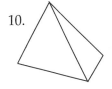 11.   12.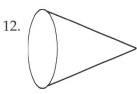

Spheres and Ovoids

	Name	Vertices	Surfaces
13.	Spheres	0	curved
14	Ellipsoids	0	curved
15.	Ovoids	0	curved

13.  14.  15.

Composite Figures

Made of two other figures or parts of figures. Look at the drawings for some examples.

16.  17. 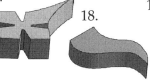 18. 19.   20.

Geometric Figures

3.26 Homework

Identify these figures. If it is a composite figure, can you identify its parts?

1.
2.
3.

4.
5.
6.

7.
8.
9.

10.
11.
12.

13.
14.
15.

16.
17.
18.

Geometric Figures

3.27 Vocabulary

    Cone
    Ellipsoid
    Identify
    Ovoid
    Polyhedron
    Prism
    Pyramid
    Rectangular prism
    Sphere
    Triangular prism

3.28 Chapter 3 Vocabulary Review

Use these words to answer the questions and problems below. Look up unfamiliar words in the dictionary. There may be more than one correct answer.

acute triangle	identify	protractor
angles	information	pyramid
base	irregular quadrilateral	quadrilateral
base line	isosceles trapezoid	rhombuses
bridges	isosceles triangle	rectangular prism
build	left	rectangle
category	length	regular
characteristic	line	right
classify	line up	right triangle
cone	matrix	scalene triangle
cross	memory chart	scale
cube	obtuse triangle	shapes
cylinder	octagon	sides
degrees	ovoid	strong
draw	parallel	solid
ellipsoid	parallelogram	sphere
equilateral	pentagons	square
figures	perpendicular	surface
geometry	point	symbol
height	polygons	triangular prism
hexagon	polyhedron	triangle
hole	prism	trapezoid
house	property	width

Geometric Figures

1. Geometry is the study of _____ .
2. Two lines do not touch. They are _____ .
3. An angle of 90° is a _____ angle.
4. Two lines are _____ if their angle is 90°.
5. We use a _____ to measure angles.
6. A protractor has a right and a left _____ .
7. We measure angles in _____ (°).
8. An _____ _____ has all angles less than 90°.
9. An _____ _____ has three sides and two are equal.
10. An _____ triangle has all sides equal.
11. An equilateral triangle is also an _____ _____ .
12. Trapezoids have two _____ parallel.
13. Parallelograms have opposite _____ equal.
14. Squares have all _____ and _____ equal.
15. _____ are parallelograms with all sides equal.
16. Regular _____ have all angles of 108°.
17. We use a protractor and ruler to _____ figures.
18. An _____ has eight angles.
19. To draw a cube first, draw a _____ .
20. You can make many _____ with cubes.
21. A rectangular _____ has eight vertices.
22. A _____ and a _____ have curved surfaces.
23. Composite _____ are made of two other _____ .

3.29 CHAPTER 3 PRACTICE TEST

Directions: Circle the correct answer.

1. A right angle is:
   a. 80°
   b. 90°
   c. 100°
   d. 110°

2. Two lines that do not touch are:
   a. perpendicular
   b. rectangular
   c. parallel
   d. characteristic

3. To measure an angle we use a:
   a. protractor
   b. pentagon
   c. shape
   d. ruler

4. An angle is made of two:
   a. cubes
   b. prisms
   c. spheres
   d. lines

Geometric Figures

5. How many equal sides are in an isosceles triangle?
    a. 0
    b. 1
    c. 2
    d. 3

6. A triangle with all small angles is:
    a. scalene
    b. acute
    c. right
    d. obtuse

7. How many parallel sides in a trapezoid?
    a. 1
    b. 2
    c. 3
    d. 4

8. A rectangle is also:
    a. a square
    b. a rhombus
    c. a polyhedron
    d. a parallelogram

9. A regular hexagon has angles of:
    a. 120°
    b. 135°
    c. 155°
    d. 168°

10. To draw a cube we start with a:
    a. triangle
    b. pentagon
    c. hexagon
    d. square

11. How many vertices in a rectangular prism?
    a. 4
    b. 6
    c. 8
    d. 10

12. Which figure has triangular surfaces?
    a. a cube
    b. a cylinder
    c. an elipseoid
    d. an icosahedron

13. Which triangle has a 90° angle?
    a. a right
    b. an acute
    c. an equilateral
    d. a scalene

14. Which figure has a circle in it?
    a. a pyramid
    b. a cone
    c. a cube
    d. a prism

Geometric Figures

3.30 Answers to Puzzles

From Section 3.1

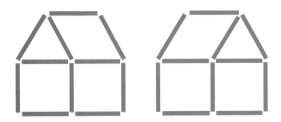

From Section 3.4
The first seven Square Numbers are 1, 4, 9, 16, 25, 36, and 49.

From Section 3.6
There are 12 rectangular prisms in the figure.

From Section 3.7
The first seven pentagonal numbers are 1, 5, 12, 22, 35, 51, and 70.

From Section 3.9
The figure has 16 small triangles, 6 with four small triangles in them, 3 with nine small triangles in them, and 1 large triangle. The total is 26 triangles in the figure.

From Section 3.21
The first seven hexagonal numbers are 1, 6, 15, 28, 47, 68, and 93.

Chapter 4

Addition & Subtraction

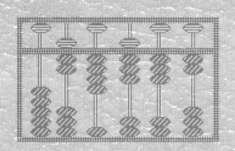

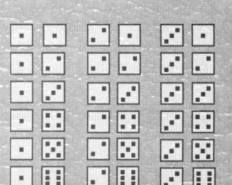

4.1 Addition and Subtraction

We write problems in two different ways. One way is horizontally. The other way is vertically.

This problem is horizontal.

$$83 + 47 = 130$$

We say, "Eighty-three plus forty-seven equals one hundred thirty."

This problem is vertical.

$$\begin{array}{r} 83 \\ + 47 \\ \hline 130 \end{array}$$

We say, "Eighty-three plus forty-seven equals one hundred thirty." We say them the same way.

This subtraction problem is horizontal.

$$193 - 76 = 117$$

We say, "One hundred ninety-three minus seventy-six equals one hundred seventeen."

This problem is vertical.

$$\begin{array}{r} 193 \\ - 76 \\ \hline 117 \end{array}$$

We say, "One hundred ninety-three minus seventy-six equals one hundred seventeen." We say them the same way.

Challenge

Remove six sticks from this figure, and leave only two squares.

Addition & Subtraction

4.2 Practice

Set A: Read these problems aloud.

1. 17 + 44 = 61
2. 59 + 83 = 142
3. 97 - 53 = 44
4. 123 - 85 = 38

5. 189
   + 84
   ───
   273

6. 217
   + 63
   ───
   280

7. 587
   + 95
   ───
   682

8. 698
   - 88
   ───
   610

9. 744
   - 29
   ───
   715

Set B: Spell out these problems with words.

10. 83 + 97 = 180
11. 15 + 19 = 34
12. 59 - 45 = 14
13. 88 - 75 = 13

14. 275
    + 97
    ───
    372

15. 458
    + 75
    ───
    533

16. 621
    + 95
    ───
    716

17. 487
    - 79
    ───
    408

18. 652
    - 38
    ───
    614

Set C: Write these problems horizontally with numerals.

19. Forty-three plus seventy-six equals one hundred nineteen
20. Eighty-four plus fifty-three equals one hundred thirty-seven
21. One hundred eighteen minus sixty-nine equals forty-nine
22. Ninety-six minus fourteen equals eighty-two

Set D: Write these problems vertically with numerals.

23. Seventy-eight plus ten equals eighty-eight
24. One hundred sixty-nine plus four hundred seventy-eight equals six hundred forty-seven
25. Five hundred eighty-one minus four hundred ninety-five equals eighty-six

4.3 Homework

Set A: Spell out these problems with words.

1. 86 + 59 = 145
2. 123 + 78 = 201
3. 457 + 119 = 576
4. 195 - 87 = 108
5. 459 - 237 = 222
6. 672 - 599 = 73

Addition & Subtraction

7. 567
   +78

   645

8. 412
   +55

   467

9. 198
   +88

   286

10. 412
    -123

    289

11. 319
    -57

    262

Set B: Write these problems with numerals. Write them horizontally.

12. Fifty-three plus seventy-nine equals one hundred thirty-two
13. Eighty-five plus one hundred fifty-three equals two hundred thirty-eight
14. One hundred eighty-four minus fifty-nine equals one hundred twenty-five
15. Two hundred ninety-two minus forty-seven equals two hundred forty-five
16. Five hundred eighty-one minus four hundred ninety-six equals eighty-five

Set C: Write these problems vertically with numerals.

17. Seventy-five plus eighteen equals ninety-three.
18. One hundred forty-two plus four hundred sixty-seven equals six hundred nine.
19. Three hundred seven minus two hundred twenty-five equals forty-seven.
20. Six hundred eighty-four minus five hundred ten equals one hundred seventy-four.
21. Two hundred eighteen minus one hundred seventy-eight equals forty.

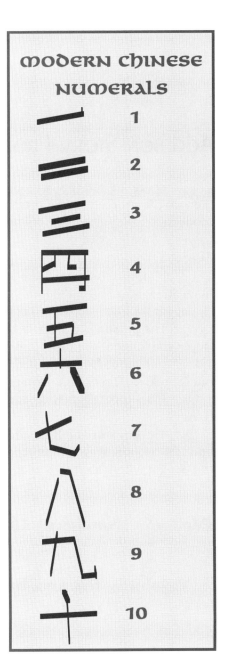

Addition & Subtraction

4.4 Vocabulary

Equals
Horizontal
Minus
Plus
Vertical

4.5 Addition Story Problems

There are many story problems. You have to read the story, and then do the problem. Most problems follow models. We will study these models.

Model 1:
John has five toys. He buys two more toys. How many toys does he have all together?

The **person** is the one who acts. A problem will also use pronouns (he or she) or proper nouns (a name).
The **numbers** count or measure things. Sometimes they are spelled out (one, two, three, etc.). Sometimes they are numerals (1, 2, 3, etc.).
The **units** tell us what is counted or measured. **Units** can be anything.
The **key words** tell us what operation to do. "All together" means to add. The numbers are in the story. The other words tell us the units and the person who is in the story.

Person   Numbers   Units   Key words

John has **five toys**. **He** buys **two *more* toys**. How many **toys** does he have ***all together***?

Addition & Subtraction

We can make many new problems by changing the person, numbers, and units.

Person/Name
    John (he)   Sam (he)   Mary (she)

Numbers
    five/two   eight/ten   seventeen/six

Units
    toys   apples   basketballs

New problem #1: Sam has eight apples. He buys ten more apples. How many apples does he have all together?

New problem #2: Mary has seventeen basketballs. She buys six more basketballs. How many basketballs does she have all together?

4.6 PRACTICE

Use Model 1. Write three new problems. Use other persons, numbers, and units.

4.7 HOMEWORK

Read these problems. List the person, numbers, units, and key words. Then solve the problem.

    Example: Sam has eight apples. He buys ten more apples. How many apples does he have all together?

Person:
    Sam (he)
Numbers:
    eight and ten
Units:
    apples
Key words:
    all together

$$\begin{array}{r} 8 \\ +10 \\ \hline 18 \end{array} \text{ apples}$$

Addition & Subtraction

1. Juan has seventeen horses. He buys twenty-nine more horses. How many horses does he have all together?
2. Maria has fifty-eight books. She buys fourteen more books. How many books does she have all together?
3. Wayne has twenty-one golf balls. He buys thirty-nine more golf balls. How many golf balls does he have all together?
4. Ricky has ninety-seven comic books. He buys one hundred eighteen more comic books. How many comic books does he have all together?
5. Samuel has four hundred eighty-five apples. He buys one hundred twenty-eight more apples. How many apples does he have all together?
6. Gerald has three hundred sixteen candy bars. He buys four hundred eighty-seven more candy bars. How many candy bars does he have all together?
7. Pamela has one thousand four hundred apples. She buys eight hundred seventy-five more apples. How many apples does she have all together?
8. Monica has six thousand one hundred seventy IBM stocks. She buys three thousand five hundred eighty-one more IBM stocks. How many IBM stocks does she have all together?
9. Jessica has forty-four thousand eight hundred twenty-five gold coins. She buys seven thousand fifty-six more gold coins. How many gold coins does she have all together?
10. Javier has two hundred eighty-seven thousand three hundred eighty-one acres of land. He buys seventy-five thousand eight hundred eleven more acres of land. How many acres of land does he have all together?

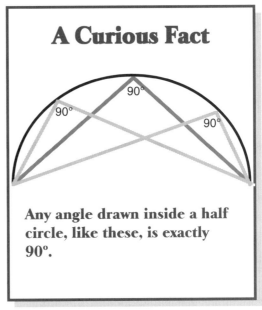

A Curious Fact

Any angle drawn inside a half circle, like these, is exactly 90°.

Addition & Subtraction

4.8 Vocabulary

All together
Key words
More
Number
Person
Story problem
Unit

4.9 Adding Money

We often add money. At the store we add prices. At home we add our bills. At the cafeteria we add up how much the food will cost.

Some key words that will help you do story problems with money are listed on this chart. Look for them in these problems.

Example 1: You buy food at the store. You come home. The register tape is torn. Can you add these prices?

The word add tells you what to do.

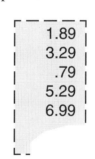

```
   1.89
   3.29
    .79
   5.29
  +6.99
 $18.25
```

Example 2: You look in your pocket. You find some coins. How much do you have?

The question, "How much do you have?" tells you what to do.

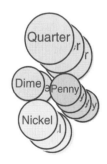

25¢ + 25¢ + 25¢ + 10¢ + 10¢ + 5¢ +
5¢ + 5¢ + 1¢ + 1¢ + 1¢ + 1¢ = $1.14

99

Addition & Subtraction

Example 3: You go to the store. You want to buy some candy. How much will it cost?

The important words are **buy** and **cost**. It tells you to add the prices.

$.79
.47
.55
+ .59
$2.40

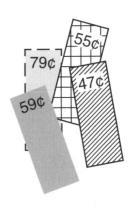

Example 4: You pay the bills. These are your checks. How much did you spend?

The **key word** is spend. It tells you to add.

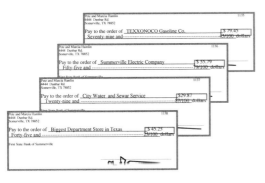

$79.45
$55.79
$29.87
+ $45.25
$210.36

4.10 Practice

Read these problems and write the person, numbers, units, and key words. Then find the answer.

1. You buy food at the store. You come home. The register tape is torn. It says $3.79, $2.99, $1.09, and $.79. Can you add these prices?
2. You look in your pocket. You find 2 quarters, 1 dime, and 5 nickles. How much do you have?
3. You go to the store. You want to buy some candy. The prices are 49¢, 79¢, and 89¢. How much will it cost?
4. You pay the bills. You write checks for $99.15, $75.00, $84.11, $100.45, and $21.16. How much did you spend?

4.11 American Money

Coins

Penny	1¢ or $.01
Nickle	5¢ or $.05
Dime	10¢ or $.10
Quarter	25¢ or $.25
Half dollar	50¢ or $.50

Bills

A dollar bill	$1.00
Five dollars	$5.00
Ten dollars	$10.00
Twenty dollars	$20.00
Fifty dollars	$50.00
One hundred dollars	$100.00

Checks

Checks, money orders, cashier's checks, certified checks, and credit card charge slips can be written for any amount. They are all forms of money.

4.12 Homework

Read these problems and write the person, numbers, units, and key words. Then find the answer.

1. You buy food at the store and come home. The register tape is torn. It says $5.99, $2.19, $6.45, $.89, and $.49. Can you add these prices?

2. You look in your pocket. You find 4 quarters, 3 dimes, 2 nickles, and 7 pennies. How much do you have?

3. You look in your pocket. You find 5 quarters, 1 dime, 12 nickles, and 21 pennies. How much do you have?

4. You go to the store. You want to buy some candy. It costs 39¢, 59¢, and 79¢. How much will it cost?

5. You pay the bills and write checks for $45.99, $75.25, $88.12, $125.00, $12.99, and $27.67. How much did you spend?

Addition & Subtraction

4.13 Vocabulary

All together
More
Cost
Add
Spend
Buy
How much do you have?

A Strange Pattern

Choose a number like 5.
Multiply it by 9.
5 x 9 = 45
Now multiply 45 by
123456789.
45 x 123456789 = 5555555505
Try any other number.
Multiply it by 9.
Then multiply the answer
times 123456789.
Do you get the same results?

4.14 Subtraction Story Problems

Many story problems use subtraction. We lose things. We eat things. We spend money. We give things to others. We sell things. All these actions can be made into subtraction problems. The **key word** for many subtraction problems is **left.**

Model 2:

Samuel buys nine pizzas. He eats three. How many does he have left?

Using this model, we can write more problems.

Samuel buys **nine pizzas**. **He** eats **three**. How many does **he** have **left?**

Person

Samuel (he)   Joseph (he)   Isabel (she)

Numbers

nine/three   sixteen/nine   twenty/six

Units

pizzas   candy bars   grapes

New problem #1: Joseph buys sixteen candy bars. He eats nine. How many does he have left?

New problem #2: Isabel buys twenty grapes. She eats six. How many does she have left?

4.15 Practice
Use Model 2 to write three new problems. Use other persons, numbers, and units.

4.16 More Subtraction Story Problems

Some subtraction problems use things to sell. Look at this model.

Model 3:
Maria sells newspapers. She has fifty newspapers to sell. She sells thirty-three. How many does she have left?

Using this model, let's write more problems.

Maria sells **newspapers. She** has **fifty newspapers** to sell. **She** sells **thirty-three.** How many does **she** have **left?**

Person
Maria/she   Jennifer/she   Matthew/he
Numbers
50/33   24/7   100/66
Units
papers   candy bars   raffle tickets

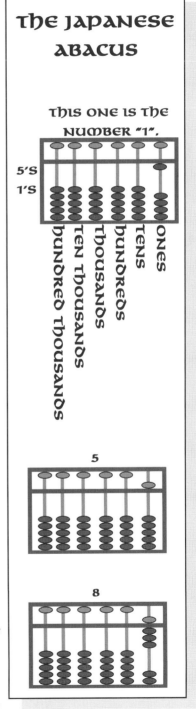

Addition & Subtraction

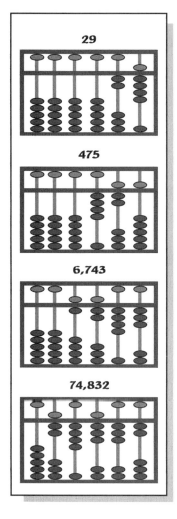

New problem #1: Jennifer sells candy bars. She has twenty-four to sell. She sells seven. How many does she have left?

New problem #2: Matthew sells raffle tickets. He has one hundred to sell. He sells sixty-six. How many does he have left?

4.17 Practice

Use Model 3 to write three new problems. Use other persons, numbers, and units.

4.18 Homework

Read these problems. List the person, numbers, units, and key words. Then solve the problem.

1. Harry buys four sandwiches. He eats two. How many does he have left?

2. Patricia buys one hundred nine candies. She eats fifty-six. How many does she have left?

3. William buys twelve pears. He eats nine. How many does he have left?

4. Caroline buys twenty-four hotdogs. She eats three. How many does she have left?

5. Jesse sells dance tickets. He has one hundred fifty to sell. He sells ninety-seven. How many does he have left?

6. Julie sells candy grams. She has seventy-five to sell. She sells thirty-nine. How many does she have left?

7. Omar sells comics. He has fifty-nine to sell. He sells twenty-five. How many does he have left?

8. Susan sells lollipops. She has one hundred forty-four to sell. She sells eighty-seven. How many does she have left?

4.19 Vocabulary

Buy
At
Give
Left
Lose
Sell
Spend

4.20 Getting Change Problems

We spend money at the store. We add the prices to see how much we spend all together. Then we subtract to find how much change we receive. This kind of problem uses addition and subtraction. The key word is change.

Some problems only subtract. Look at this model. The key word is "change." It tells you to subtract.

Model 4:
Mary has $20.00. She spends $12.95 at the grocery store. How much **change** does she receive?

Person
  Mary/she; Henry/he; Gabriel/he
Numbers
  20.00/12.95 50.00/33.50 10.00/7.00
Units
  These stay the same in change problems. $ $ $
Kind of store
  grocery, auto parts, drugstore

Pairs of Number Cubes

This figure shows all the combinations of two number cubes. Look at the combinations. How many add up to 6? How many add up to 7? How many add up to 12?

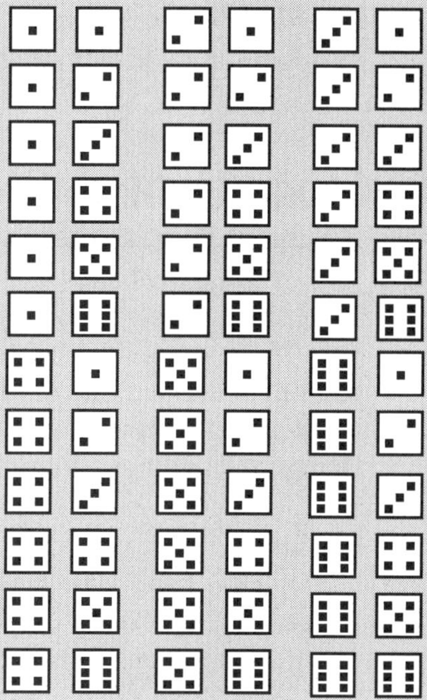

Make a list.

2 comes up _____ time(s).
3 comes up _____ time(s).
4 comes up _____ time(s).
5 comes up _____ time(s).
6 comes up _____ time(s).
7 comes up _____ time(s).
8 comes up _____ time(s).
9 comes up _____ time(s).
10 comes up _____ time(s).
11 comes up _____ time(s).
12 comes up _____ time(s).

Addition & Subtraction

New problem #1: Henry has $50.00. He spends $33.50 at the auto parts store. How much change does he receive?

New problem #2: Gabriel has $10.00. He spends $7.00 at the drug store. How much change does he receive?

4.21 PRACTICE

Use Model 4 to write two new problems. Use other persons, numbers, and kinds of stores.

4.22 MORE GETTING CHANGE PROBLEMS

Some problems use addition and subtraction. First, you add all the things to buy. Then, you subtract from the money you have. Look at this model. The key word **and** tells you to add. The key word **change** tells you to subtract.

> The key word **and** tells you to add.
> $1.25
> +.79
> $2.04
>
> Then the key word **change** tells you to subtract
> $5.00
> -2.04
> $2.96

Model 5:
Hiro has $5.00. He buys a hamburger for $1.25 and a drink for $.79. How much change does he receive?

Person
Hiro he	Alice she	Daphne she

Numbers
$5.00/$1.25/$.79	$20.00/$4.95/$2.95	$10.00/$2.95/$3.15

Units
$	$	$

Things to buy
hamburger/drink	book/magazine	lipstick/mascara

Addition & Subtraction

New problem #1: Alice has $20.00. She buys a book for $4.95 and a magazine for $2.95. How much change does she receive?

New problem #2: Daphne has $10.00. She buys a lipstick for $2.95 and a mascara for $3.15. How much change does she receive?

4.23 Practice

Write two new problems. Use other persons, numbers, and things to buy.

4.24 Homework

Read these problems. List the person, numbers, units, and kind of store or things to buy. Then solve the problem.

1. Juan has $30.00. He spends $23.90 at the discount store. How much change does he receive?

2. Phil has $20.00. He buys a fan belt for $5.95 and a heater hose for $7.25. How much change does he receive?

3. Li Mei has $35.00. She buys a textbook for $22.95 and a notebook for $1.95. How much change does she receive?

4.25 Vocabulary

All together
More
Cost
Add
Spend
Buy
How much do you have?
Left
Change

Addition & Subtraction

4.26 Mixed Story Problems

Most mathematics books have many story problems. They have many kinds on the same page. Sometimes you will have to add. Sometimes you will have to subtract or multiply. Always look at the key words for help. This list of key words tells you the operation to use.

all together +
more +
cost +
add +
spend +
buy +
how much +
left -
change - or + & -
both +
all +
each x

4.27 Homework

Read these problems. Underline the key words. Then do the problems.

1. Roberto has fifteen baseballs. He buys five more baseballs. How many baseballs does he have all together?
2. Michael has $30.00. He spends $24.95. How much change does he get?
3. Vishnu paid his bills. They were $55.99, $84.75, and $125.76. How much are all the bills?
4. Mr. Park owns three stores. He has fifteen employees at one store, twenty-eight at the second store, and nineteen at the third store. How many employees does he have all together?
5. Charles had twenty-nine students. He lost three last week. How many does he have left?
6. Tina has five hundred patches. She buys twenty-five more. How many does she have all together?
7. Alicia has a ten dollar bill, five quarters, and two pennies. How much money does she have?
8. Alfred has $20.00. He buys a drink for $.89, a hotdog for $1.29, and a bag of chips for $.79. How much change does he receive?
9. Paula wants to buy a camera for $179.88 and a flash attachment for $89.99. How much will both cost?

Magic Triangle

Put the numbers 1, 2, 3, 4, 5, and 6 in the circle. Each side must add up to ten.

10

Addition & Subtraction

10. Alex has $20.00. He spends $12.50. How much change does he receive?
11. David has eighteen football stickers. He buys three more. How many stickers does he have?
12. Susan drove 58.7 miles on Monday and 78.3 miles on Tuesday. How many miles did she drive on both days?
13. Dora has $10.00. She buys three candy bars. They cost 79¢ each. How much change does she receive?
14. Frank has $10.00. He buys an order of fries for $.89, a hamburger for $1.89, and a drink for $.69. How much change does he get?
15. Carmen wrote checks for $18.99, $43.77, and $78.00. How much are all the checks?
16. Hue has eighty watches to sell. She sells fourteen. How many are left?
17. Jeanie has fifty-five silver thimbles. She buys six more on vacation. How many thimbles does she have?
18. LaToya had seven quarters. She lost two. How much money does she have left?
19. Larry buys three magazines. They cost $3.95, $2.95, and $4.50. How much do all the magazines cost?
20. Belinda wants to buy a candy bar for $.89, a magazine for $2.95, a burrito for $1.29, a drink for $.75, and an ice cream bar for $.79. How much will it cost all together?
21. John drives 2.5 miles to work. He drives 4.2 miles to the store and then 3.9 miles back home. How many miles does he drive all together?
22. Leland paid his bills. He paid $55.75 to the electric company, $95.00 to the bank, $250.60 for his car, $798.00 for the mortgage, and $45.33 for the water bill. How much did he pay in all?
23. Mario has fifteen apples. He eats six. How many does he have left?
24. Olga had five keys. She lost two. How many does she have left?

Another Strange Pattern

Multiply 2 x 7 = 14
Then multiply 14 x 15873.
The answer is 222222.

Try 3 x 7 = 21
Then 21 x 15873 =

What is the answer?

Try other numbers.

Addition & Subtraction

25. Mr. Apodaca teaches science. He has twenty-three students in third period class and twenty-nine students in fifth period. How many students are in both classes?
26. Fatima had $50.00. She gave $20.00 to her son. How much does she have left?
27. Karen buys 2.5 pounds of ham and 3.2 pounds of turkey at the delicatessen. How much do both packages weigh?
28. Taylor bought four books. They cost $12.95, $21.50, $8.95, and $15.00. How much did they all cost?
29. Austin wants to buy a VCR for $219.99, and a TV for $289.99. How much will both cost?
30. Ana Maria has a five dollar bill, three quarters, and two dimes. How much money does she have?
31. Dean has one hundred sixteen cars to sell. He sells sixty-one. How many cars are left?

4.28 Chapter 4 Vocabulary Review

Use these words to answer the questions and problems below. Look up unfamiliar words in the dictionary. There might be more than one correct answer.

action	identify	price
add	key words	problems
all together	kinds	sell
bill	left	solve
buy	list	store
change	lose	story
chart	means	spend
checks	model	tell
cost	money	things
eat	minus	toys
give	operation (s)	units
help	person	vertical
horizontal	plus	

1. The key word, **more**, means to _____.
2. Addition and subtraction are _____.
3. Problems can be horizontal or _____.
4. The key word, _____, means to subtract.
5. We read the sign, "-", with the word, _____.

110

Addition & Subtraction

6. Quarters, dimes, and nickles are _____.
7. We write _____ to pay bills.
8. Pizzas and money can be ____ in story problems.
9. We _____ money to buy things.
10. We receive _____ at the store after we pay for something.

Unscramble these vocabulary words:

11. suimn
12. ntrpoeaio
13. puls
14. syto
15. ivge
16. inydtfe
17. khescc
18. loes
19. dad
20. sleov
21. tocnia
22. libl
23. sryto
24. eghnac
25. lvtiaerc

4.29 CHAPTER 4 PRACTICE TEST

Directions: Circle the correct answer.

1. What form is the problem, 17 + 23 = 40, written in?
   a. vertical
   b. identify
   c. all together
   d. horizontal

2. How do you say the word for the symbol "-"?
   a. unit
   b. cost
   c. plus
   d. minus

3. What operation does the key word, "more," tell you to do?
   a. multiplication
   b. subtraction
   c. addition
   d. division

4. Hamid has thirteen candy bars. He buys six more. How many does he have?
   a. 19 candy bars
   b. 13 candy bars
   c. 7 candy bars
   d. 6 candy bars

5. You look in your pocket and find two quarters, three dimes, and two pennies. How much do you have?
   a. $ .75
   b. $ .82
   c. $1.05
   d. $1.07

6. You buy some candy for 59¢, 69¢, and 25¢. How much will it cost?
   a. $1.53
   b. $1.50
   c. $ .69
   d. $ .59

Addition & Subtraction

7. Manny buys seven apples. He eats four. How many are left?
   a. 3 apples
   b. 4 apples
   c. 7 apples
   d. 11 apples

8. Ms. Brown has thirty students in one class and thirty-five in another. How many does she have in both classes?
   a. 65 students
   b. 55 students
   c. 7 students
   d. 5 students

9. Receiving change in a story problem means to:
   a. add
   b. subtract
   c. multiply
   d. divide

10. Maude has $30.00. She buys a hair blower for $14.95 and a bottle of make up for $7.99. How much change does she get back?
    a. $6.96
    b. $7.06
    c. $7.99
    d. $9.00

11. Harry buys 3.5 lbs. of oranges and 2.9 lbs. of apples. How much does his bag of fruit weigh?
    a. 0.6 lbs.
    b. 3.5 lbs.
    c. 5.4 lbs.
    d. 6.4 lbs.

12. What key word tells you to subtract?
    a. more
    b. buy
    c. left
    d. cost

4.30 Answers to Puzzles

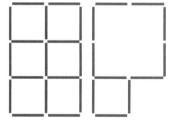

From Section 4.1
Remove six sticks and leave only two squares.

From Section 4.13

1 x 9 x 123456789 = 1111111101
2 x 9 x 123456789 = 2222222202
3 x 9 x 123456789 = 3333333303
4 x 9 x 123456789 = 4444444404
5 x 9 x 123456789 = 5555555505
6 x 9 x 123456789 = 6666666606
7 x 9 x 123456789 = 7777777707
8 x 9 x 123456789 = 8888888808
9 x 9 x 123456789 = 9999999909

Addition & Subtraction

From Section 4.19

2 comes up 1 time.
3 comes up 2 times.
4 comes up 3 times.
5 comes up 4 times.
6 comes up 5 times.

7 comes up 6 times.
8 comes up 5 times.
9 comes up 4 times.
10 comes up 3 times.
11 comes up 2 times.
12 comes up 1 time.

From Section 4.27

1 x 7 x 15873 = 111111
2 x 7 x 15873 = 222222
3 x 7 x 15873 = 333333
4 x 7 x 15873 = 444444
5 x 7 x 15873 = 555555
6 x 7 x 15873 = 666666
7 x 7 x 15873 = 777777
8 x 7 x 15873 = 888888
9 x 7 x 15873 = 999999

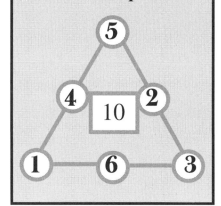

Magic Triangle

Put the numbers 1, 2, 3, 4, 5, and 6 in the circle. Each side must add up to ten.

From Section 4.27

Chapter 5

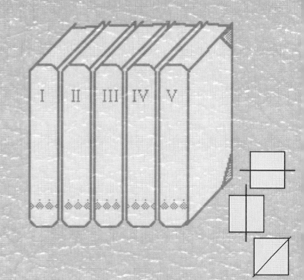

Rounding & Estimating

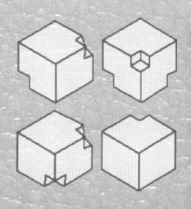

Rounding and Estimating

5.1 Rounding Tens

Rounding is approximating a number. It is changing a hard number into an easy number with a zero at the end. Rounding makes some problems easy to do.

The directions always say, "Round to the nearest ten." The **key words** are **round** and **nearest ten.**

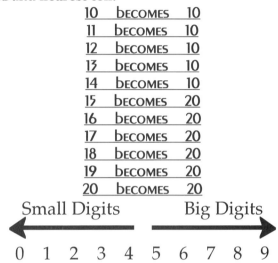

The rule for rounding tens:
First, look at the digits in the tens and ones places.

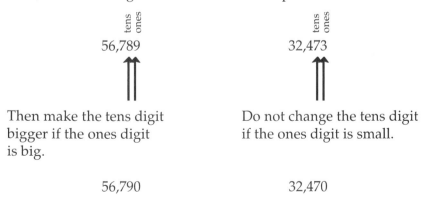

Then make the tens digit   Do not change the tens digit
bigger if the ones digit   if the ones digit is small.
is big.

        56,790   32,470

Also, the ones digit always becomes zero!

Rounding and Estimating

Move four lines and make only three squares. All the squares will be the same size. Use all the lines, but none of the sides will touch.

5.2 Examples

Round to the nearest ten.

9	becomes	10
4	becomes	0
37	becomes	40
22	becomes	20
98	becomes	100
54	becomes	50
225	becomes	230
254	becomes	250
279	becomes	280
393	becomes	390
589	becomes	590
580	becomes	580
995	becomes	1,000
994	becomes	990
4,785	becomes	4,790
5,972	becomes	5,970

5.3 Practice

Set A: Identify as big or small ones digit.

1. 5
2. 3
3. 7
4. 1
5. 2
6. 9
7. 0
8. 4
9. 6
10. 8
11. 23
12. 98
13. 37
14. 82
15. 70
16. 44

Set B: Round to the nearest ten.

17. 21
18. 43
19. 55
20. 26
21. 32
22. 60
23. 39
24. 54
25. 47
26. 78
27. 155
28. 510
29. 233
30. 664
31. 446
32. 982
33. 321
34. 707
35. 878
36. 2,199

5.4 Homework

Round to the nearest ten.

1. 14
2. 63
3. 9
4. 26
5. 41
6. 53
7. 72
8. 37
9. 85
10. 98
11. 241
12. 826
13. 482
14. 168
15. 604
16. 1,313
17. 7,757
18. 3,990
19. 9,535
20. 5,079

5.5 Vocabulary

Approximating
Bigger
Changing
Digit
Easy
End
Hard
Round
Nearest
Nearest ten

5.6 Rounding Big Numbers

How many students are in your class? We sometimes say, "About twenty." How many students are in your school? We sometimes say, "About one thousand." The word **about** means that we rounded the number.

The directions always say, "Round to the nearest _____." The key word in the blank tells you how many zeroes will be in the answer.

Rounding and Estimating

ADDING ROMAN NUMERALS

ROMAN NUMERALS WERE VERY HARD TO WORK WITH. ANY PROBLEMS COULD BE VERY LONG AND COMPLICATED. LOOK AT THE ADDITION PROBLEM.

XIX
XVIII
XXXVII
+ LXXXVIII

LXXXXXXXXIXVVVIIIIIII
THIS BECOMES...
LXXXXXXXXIXVVVVIII
THIS BECOMES...
LXXXXXXXXXXIXIII
THIS BECOMES...
LLLIXIII
THIS BECOMES...
CLIXIII
THIS BECOMES...
CLXII

THIS IS THE ANSWER. COMPARE THIS TO THE WAY WE ADD...

$$\begin{array}{r} \overset{3}{1}9 \\ 18 \\ 37 \\ +\ 88 \\ \hline 162 \end{array}$$

Rounding and Estimating

Round to the nearest **ten** means one zero in the answer.
Answers like: 10 or 20 or 140

Round to the nearest **hundred** means two zeros in the answer.
Answers like: 100 or 700 or 3200

Round to the nearest **thousand** means three zeros in the answer.
Answers like: 2,000 or 5,000 or 11,000

Round to the nearest **ten thousand** means four zeros in the answer.
Answers like: 50,000 or 90,000 or 170,000

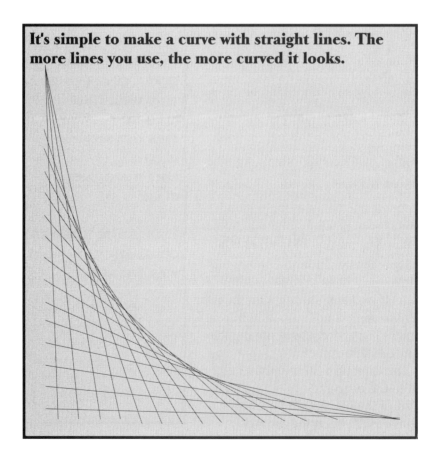

It's simple to make a curve with straight lines. The more lines you use, the more curved it looks.

Rounding and Estimating

5.7 Examples

Round to the nearest ten.
    23 becomes 20   49 becomes 50
    282 becomes 280   5,239 becomes 5,240

Round to the nearest hundred.
    214 becomes 200   386 becomes 400
    2,112 becomes 2,100   5,672 becomes 5,700

Round to the nearest thousand.
    24,723 becomes 25,000   189,534 becomes 190,000

Round to the nearest ten thousand.
    147,147 becomes 150,000   281,520 becomes 280,000

5.8 Practice

Set A: Round to the nearest ten.
1. 15   2. 60   3. 27   4. 84   5. 39

Set B: Round to the nearest hundred.
6. 122   8. 839   10. 774   12. 357   14. 109
7. 508   9. 243   11. 458   13. 660   15. 916

Set C: Round to the nearest thousand.
16. 5,151   21. 49,147
17. 6,422   22. 1,475
18. 83,898   23. 7,636
19. 9,581   24. 3,873
20. 22,006

Set D: Round to the nearest ten thousand.
25. 37,404
26. 97,810
27. 261,685
28. 85,054
29. 10,866
30. 217,477
31. 44,522
32. 39,231
33. 539,968

How big will it get?

$$1/2$$
$$+ 1/4 \; (2 \times 2 = 4)$$
$$+ 1/8 \; (4 \times 2 = 8)$$
$$+ 1/16 \; (8 \times 2 = 16)$$
$$+ 1/32 \; (16 \times 2 = 32)$$
$$+ 1/64...$$

If we continue adding more fractions like these, will the sum ever be larger than one? Maybe this graphic representation might help you decide.

Rounding and Estimating

Königsberg Bridge Problem

This map represents a city on two islands in a river. Can you cross all the bridges just one time? The bridges are labeled a through g.

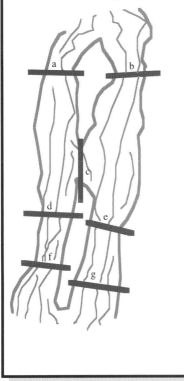

5.9 Homework

Set A: Round to the nearest ten.

1. 94
2. 30
3. 16
4. 46
5. 62
6. 70
7. 85
8. 383
9. 495
10. 674

Set B: Round to the nearest hundred.

11. 760
12. 298
13. 825
14. 133
15. 147
16. 909
17. 556
18. 704
19. 111
20. 8,780

Set C: Round to the nearest thousand.

21. 4,899
22. 9,518
23. 6,761
24. 2,013
25. 3,241
26. 2,650
27. 2,038
28. 23,092
29. 55,458

Set D: Round to the nearest ten thousand.

30. 83,770
31. 12,193
32. 46,289
33. 87,162
34. 96,004
35. 35,849
36. 149,531
37. 933,162
38. 467,570

5.10 Vocabulary

Round
Round nearest ten
Round nearest hundred
Round nearest thousand
Round nearest ten thousand

Rounding and Estimating

5.11 Estimating

We use rounding to make some problems easier. Estimating is approximating the answer to a problem. We can estimate addition problems (+). We can estimate subtraction problems (-) and multiplication problems (x). We can even estimate division problems (÷). The key word is always **estimate**.

Examples

Estimate the answers.
299 + 807 is about 300 + 800, which is 1,100.
637 - 479 is about 600 - 500, which is 100.
27 x 79 is about 30 x 80, which is 2,400.
597 ÷ 22 is about 600 ÷ 20, which is 30.

Estimating an answer is easy to do. First, round to the biggest place in the smaller number. Then do the operation. Many times you can do this in your head. It is very easy way to do a problem.

5.12 Examples

Estimate 2,243 + 793
The problem becomes: 2,200 + 800 = 3,000

Estimate 227 - 43
The problem becomes: 230 - 40 = 190

Estimate 23,475 x 2,356
The problem becomes: 23,000 x 2,000 = 46,000,000

Estimate 275,823 ÷ 123
The problem becomes: 275,800 ÷ 100 = 2,758

Round to the biggest place of the smaller number.

5.13 Practice

Estimate the answer:

1. 155 + 326 =
2. 314 + 53 =
3. 206 + 937 =
4. 4,107 + 214 =
5. 414 - 448 =
6. 9,306 - 623 =
7. 5,928 - 593 =
8. 684 - 328 =
9. 224 x 63 =
10. 1395 x 417 =
11. 7,295 x 771 =
12. 848 x 813 =
13. 6,763 ÷ 193 =
14. 5,619 ÷ 695 =
15. 7,812 ÷ 178 =

Rounding and Estimating

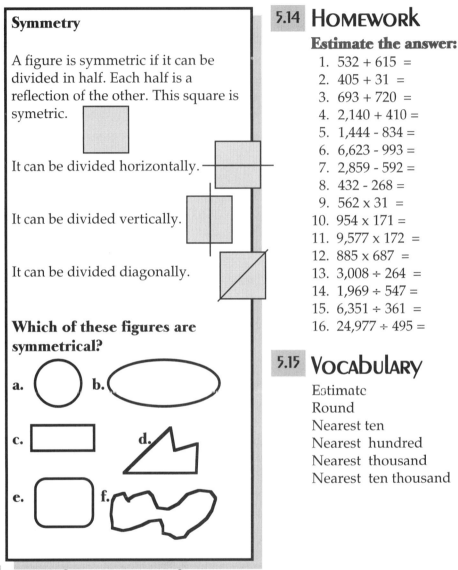

Symmetry

A figure is symmetric if it can be divided in half. Each half is a reflection of the other. This square is symetric.

It can be divided horizontally.

It can be divided vertically.

It can be divided diagonally.

Which of these figures are symmetrical?

a. b.

c. d.

e. f.

5.14 Homework
Estimate the answer:
1. 532 + 615 =
2. 405 + 31 =
3. 693 + 720 =
4. 2,140 + 410 =
5. 1,444 - 834 =
6. 6,623 - 993 =
7. 2,859 - 592 =
8. 432 - 268 =
9. 562 x 31 =
10. 954 x 171 =
11. 9,577 x 172 =
12. 885 x 687 =
13. 3,008 ÷ 264 =
14. 1,969 ÷ 547 =
15. 6,351 ÷ 361 =
16. 24,977 ÷ 495 =

5.15 Vocabulary
Estimate
Round
Nearest ten
Nearest hundred
Nearest thousand
Nearest ten thousand

5.16 Rounding Decimal Fractions

We round decimal fractions often. Sometimes we do this when we divide or multiply. We always do this when we make calculations with money. We round decimal fractions in percent problems, also.

The directions always say, "Round to the nearest _____." Then, they tell you the place name. This key word tells you how many places after the decimal point will have numbers. If the answer will have no digits after the decimal point the key words are whole numbers.

Rounding and Estimating

5.17 Examples

Round to the nearest tenth. (Each answer has only one decimal place.)
12.88 becomes 12.9
17.01 becomes 17.0
24.953 becomes 25.0

Round to the nearest hundredth. (Each answer has two decimal places.)
5.123 becomes 5.12
12.467 becomes 12.47
31.091 becomes 31.09

Round to the nearest thousandth. (Each answer has three decimal places.)
4.6432 becomes 4.643
0.9659 becomes 0.966
93.00461 becomes 93.005

Round to the nearest whole number. (Each answer is a whole number.)
16.43 becomes 16
29.5 becomes 30
67.49 becomes 67
0.21 becomes 0
0.556 becomes 1

5.18 Practice

Set A: Round to the nearest tenth.
1. 16.87
2. 93.156
3. 27.36
4. 21.05

Set B: Round to the nearest hundredth.
5. 6.423
6. 4.611
7. 3.5442
8. 0.325

Set C: Round to the nearest thousandth.
9. 1.4314
10. 4.8253
11. 3.6422
12. 1.5436

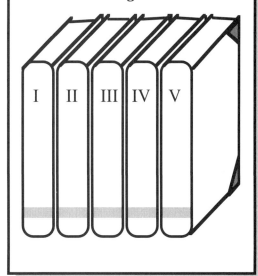

The Bookworm

A bookworm began eating his way through a set of five books on the shelf. He began at the front cover of volume one and ate through to the back cover of volume five. If each volume was one inch thick, how many inches did he eat through?

Rounding and Estimating

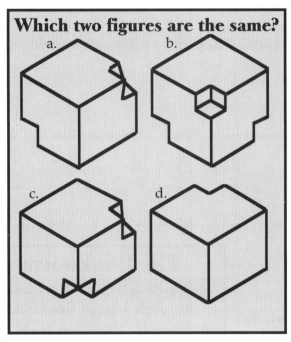

Which two figures are the same?

Set D: Round to the nearest whole number.
13. 4.4
14. 3.31
15. 16.27
16. 25.6

5.19 Homework

Set A: Round to the nearest tenth.
1. 27.78
2. 56.562
3. 16.49
4. 21.06
5. 3.42
6. 93.33
7. 42.821
8. 69.98

Set B: Round to the nearest hundredth.
9. 6.144
10. 3.215
11. 12.015
12. 9.995

Set C: Round to the nearest thousandth.
13. 0.5624
14. 16.0255
15. 5.9236
16. 23.9996

Set D: Round to the nearest whole number.
17. 0.6
18. 24.12
19. 6.99
20. 12.45

5.20 Vocabulary

Nearest tenth
Nearest hundredth
Nearest thousandth
Nearest whole number

Rounding and Estimating

5.21 Rounding in Money Problems

Our money comes in dollars and cents. Cents are decimal fractions. Cents are hundredths of a dollar.

One dollar is written "$1.00." Twenty-five cents can be written either "25¢" or "$.25."

Since $.01 is the smallest amount of money we have, we have to round off to the nearest cent in many calculations.

5.22 Examples

```
      $1.25
    x  .25
       605
      2500
    $.3105   becomes  $.31
```

```
       $ .678  becomes $.68
    7 )$4.75
        4 2
          55
          49
          60
          56
           4...
```

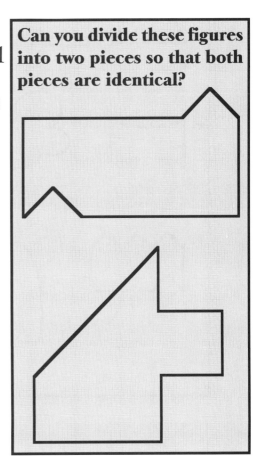

Can you divide these figures into two pieces so that both pieces are identical?

"Round off to the nearest cent," means, "Round off to the nearest hundredth." We use the same rule. We put a dollar sign ($) in front of the number.

Rounding and Estimating

Symmetry Again!
Which of these figures are symmetrical?

a.

b.

c.

d.

e.

f.

g.

h.

5.23 Practice
Round off to the nearest cent.
1. $6.1151
2. $52.978
3. $82.154
4. $.7469
5. $.3472
6. $36.568
7. $10.521
8. $65.235
9. $.3893
10. $96.286
11. $23.014
12. $78.077
13. $47.863
14. $.422
15. $.69345
16. $.0098

5.24 Homework
Set A: Round off to the nearest cent.
1. $2.101
2. $5.576
3. $10.829
4. $.2421
5. $4.317
6. $.4383
7. $7.784
8. $.6552

Set B: Multiply and then round off to the nearest cent.
9. $2.35 x 6.22 =
10. $12.33 x 6.2 =
11. $15.90 x 0.52 =
12. $2.25 x 0.1 =
13. $14.23 x 0.16 =
14. $19.99 x 3.5 =

Rounding and Estimating

Set C: Divide and then round off to the nearest cent.
15. $15.02 ÷ 6 =
16. $3.25 ÷ 8 =
17. $5.50 ÷ 8 =
18. $2.50 ÷ 4 =
19. $100.00 ÷ 13 =
20. $200.00 ÷ 3 =

5.25 Vocabulary

Dollar
Dollar sign
Cent
Money
Nearest cent

5.26 Chapter 5 Vocabulary Review

Use these words to answer the questions and problems below. Look up unfamiliar words in the dictionary. There might be more than one correct answer.

about	fractions	front	round off
aproximating	decimal	has	rounding
addition	places	head	rule
always	decimal	hundred	same
answer	point	hundredth	say
amount	directions	like	since
becomes	divide	money	small
big	division	multiplication	smallest
biggest	dollar sign	multiply	subtraction
both	dollars	nearest	ten
calculation	easier	often	ten thousand
cent	easy	ones place	tenth
change	either	percent	thousand
close	estimate	problem	thousandth
comes	estimating	raise	whole number
decimal	estimation	round	very

1. Round to the _____ ten.
2. The number 5.43 has two _____ places.
3. 8 + 4 is an _____ problem.
4. For 12 · 3 = 36, 36 is the _____ .
5. In the number 8.53, the symbol "." is a _____ _____ .
6. The symbol "$" is the _____ .
7. _____ is approximating the value of a number.
8. The number 84 _____ 80 when you round.
9. The number $3.988 needs to be rounded to the nearest _____ .
10. The problem 800 + 700 is an _____ of the problem 823 + 698.

129

Rounding and Estimating

5.27 Chapter 5 Practice Test

Directions: Circle the correct answer.

Round to the nearest ten:

1. 84
   a. 70
   b. 80
   c. 85
   d. 85

2. 23
   a. 10
   b. 20
   c. 25
   d. 30

3. 115
   a. 105
   b. 110
   c. 115
   d. 120

Round to the nearest hundred:

4. 224
   a. 100
   b. 120
   c. 130
   d. 200

5. 569
   a. 500
   b. 560
   c. 570
   d. 600

6. 1,459
   a. 1,000
   b. 1,400
   c. 1,460
   d. 1,500

Round to the nearest tenth:

7. 17.45
   a. 17
   b. 17.4
   c. 17.5
   d. 18

8. 2.31
   a. 2
   b. 2.3
   c. 2.4
   d. 3

9. 0.821
   a. 0.7
   b. 0.8
   c. 0.82
   d. 1.0

Round to the nearest cent:

10. $8.215
    a. $8.00
    b. $8.21
    c. $8.22
    d. $8.30

11. $.123
    a. $.11
    b. $.12
    c. $.13
    d. $.20

12. $12.4379
    a. $12.40
    b. $12.43
    c. $12.44
    d. $12.438

Estimate the answer:

13. 88 + 19
    a. 100
    b. 107
    c. 110
    d. 117

14. 262 - 37
    a. 220
    b. 225
    c. 230
    d. 260

From Section 5.22

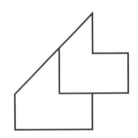

Rounding and Estimating

5.28 ANSWERS TO PUZZLES

From Section 5.2

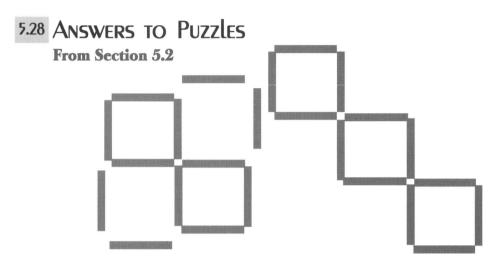

From Section 5.8

The addition of the fractions 1/2 + 1/4 + 1/8 + 1/16 ... will never be bigger than the number one. The reason is that you always go just half the distance that remains.

From Section 5.9

It is impossible to cross all the bridges just one time. The reason for this is the odd number of bridges that lead to the northern island.

From Section 5.14

The figures a, b, c, and e are symmetrical.

From Section 5.17

The book worm would eat through three inches only.

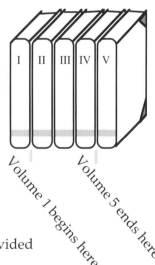

From Section 5.19

Figures B and C could be the same. To see this, rotate figure B one quarter turn in a counter-clockwise direction.

From Section 5.22

The figure on the opposite page can be divided into two equal parts. The other cannot.

From Section 5.23

Figures a, c, e, and h are symmetrical.

Chapter 6

Multiplication

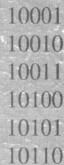

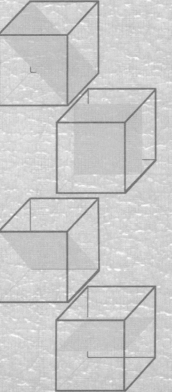

10000
10001
10010
10011
10100
10101
10110
10111
11000
11001
11010
11011
11100
11101
11110
11111
100000
100001
100010
100011
100100
100101
100110
100111
101000
101001
101010
101011
101100
101101
101110

Multiplication

6.1 Multiplication

We write multiplication problems in two different ways. One way is horizontally. The other way is vertically.

This problem is horizontal.

$$593 \times 127 = 75{,}311$$

We say, "Five hundred ninety-three times one hundred twenty-seven equals seventy-five thousand three hundred eleven." The word times always means to multiply.

This problem is vertical.

$$\begin{array}{r} 593 \\ \times\,127 \\ \hline 75{,}311 \end{array}$$

We say, "Five hundred ninety-three times one hundred twenty-seven equals seventy-five thousand three hundred eleven." We say it the same way as the first problem.

6.2 Practice

Read these problems aloud, then spell them out with words.

1. 48 x 63 = 3,024
2. 189 x 116 = 21,924
3. 648 x 96 = 62,208
4. 142 x 93 = 13,206

5.   47
  x 17
   799

6.   69
  x 83
 5,727

7.   832
 x 212
176,384

8.   1,480
   x 692
1,024,160

Tessellations

Tessellations are patterns made up of small shapes that are all the same. Some tessellations use simple shapes.

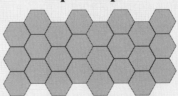

Hexagon   **Tessellation with hexagons**

Multiplication

Multiplication uses many symbols. All of these mean to multiply six and four together.

6 x 4   **6 · 4**   **6 * 4**

6(4)   **(6)4**   **(6)(4)**

Multiplication is commutative. This means that the numbers being multiplied can be moved. All these problems are the same.

73 · 42 = 3,066   **42 · 73 = 3,066**

```
    42            73
  x 73          x 42
  3,066         3,066
```

6.3 Homework

Set A: Spell out these problems with words.

1. 178(314) = 55,892
2. (269)490 = 131,810
3. 350 * 682 = 238,700
4. (412)(573) = 236,076

```
5.    83        7.   193
    x 64            x 24
    5,312           4,632

6.    79        8.   682
    x 19           x 419
    1,501         285,758
```

Remove three sticks so that only three squares are left. All three squares will be the same size.

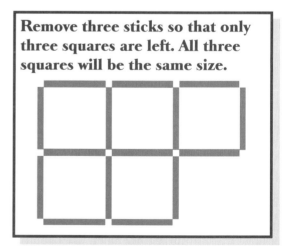

136

Multiplication

Set B: Solve these problems.

9. 12 x 25 =
10. (53)(4) =
11. 62(4) =
12. 17 x 6 =
13. 11 x 9 =
14. 31 · 14 =
15. 5 · 8 =
16. 10 * 14 =
17. 16 x 27 =
18. (26)13 =
19. 13 x 26 =
20. 6(17) =
21. 21 · 37 =
22. 11 * 19 =
23. 75(11) =
24. 14 x 10 =
25. (25)(12) =
26. 49 * 11 =
27. 42 * 16 =
28. 23 · 6 =
29. (9)(11) =
30. 18 x 5 =
31. 15 x 7 =
32. 32 · 20 =
33. (7)(15) =
34. 27(16) =
35. 19 x 11 =
36. 22 . 11 =
37. 45 * 10 =
38. 65(14) =
39. (51)(9) =
40. (59)6 =

6.4 Extension

Look at the problems above. List all the pairs of problems that are the same. These show the commutative property.

Example: 45 * 17 and 17(45) are the same problem.

6.5 Vocabulary

Commutative
Horizontal
List
Moved
Multiplication
Pair
Property
Symbol
Times
Vertical

Symmetrical Letters
Which letters of the alphabet are symmetrical.

A
B
C
D
E
F
G
H
I
J
K
L
M

N
O
P
Q
R
S
T
U
V
W
X
Y
Z

Multiplication

6.6 Multiplication Story Problems

Many story problems use multiplication. You have to read the story and then do the problem. Most problems follow models. We will look at several models.

Model 1

John has five books. Each one has 125 pages. How many pages are in the books?

The key word tells us what to do. **Each** tells us to multiply. The other words tell us the person, numbers, and units.

In this model we see two units: books and pages. Pages are in books. The units must go together.

Person   **Numbers**   **Units**   **Key Words**

John has five books. Each one has 125 pages. How many pages are in the books?

We can make many new problems by changing the person, numbers, and units of this model.

Person
    John   William   Susan

Numbers
    5/125   7/48   19/265

Units
    Books/pages   Bottles of soda/ounces Bags of peanuts/peanuts

New problem #1:
William has seven bottles of soda. Each one has forty-eight ounces of soda. How many ounces are in the bottles?

New problem #2:
Susan has nineteen bags of peanuts. Each one has 265 peanuts. How many peanuts are in the bags?

6.7 Practice

Write three new problems with this model. Use other persons, numbers, and units.

6.8 Homework

Read these problems. List the person, numbers, and units. Then solve the problem.

Example: Maria has 24 boxes of candy. Each one has 88 pieces of candy in it. How many pieces of candy are in the boxes?
Person: Maria
Numbers: 24 and 88
Units: boxes of candy and pieces of candy
Key words: Each

```
    24
   x 88
  2,112   pieces of candy
```

1. Ella has three bags of marbles. Each one has 89 marbles. How many marbles are in the bags?
2. Ben has nine boxes of pencils. Each one has 24 pencils. How many pencils are in the boxes?
3. Alberto has six cans of cola. Each one has twelve ounces. How many ounces are in the cans?
4. Maria has seven packages of bubble gum. Each one has six pieces of gum. How many pieces of gum are in the package?
5. Ricky has six rolls of quarters. Each one has 50 quarters. How many quarters are in the rolls?

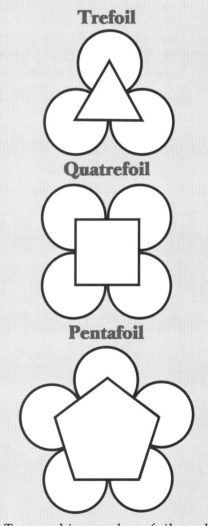

Multifoils

Multifoils are regular polygons with arcs of circles around each vertex. They are called trefoils, quatrefoils, pentafoils, hexafoils, octafoils, etc.

Trefoil

Quatrefoil

Pentafoil

Try making a hexafoil and octafoil for yourself.

Multiplication

6. Juanita has nine boxes of paper clips. Each one has fifty paper clips. How many paper clips are in the boxes?
7. Donald has three packages of notebook paper. Each one has 250 sheets of paper. How many sheets are in the packages?
8. Harry has forty-nine packages of ballpoint pens. Each one has three pens. How many pens are in the packages?
9. April has twelve bags of rubber bands. Each one has 100 rubber bands. How many rubber bands are in the bags?
10. Juan has nineteen jigsaw puzzles. Each one has 150 pieces. How many pieces are in the puzzles?
11. Patricia has three sets of dominoes. Each one has forty-eight dominoes. How many dominoes are in the sets?
12. Jim has six bags of hamburgers. Each one has four hamburgers. How many hamburgers are in the bags?

6.9 Vocabulary

Bag
Book
Bottle
Box
Change
Each
Model
Number
Page
Peanut
Person
Several
Story
Times
Units

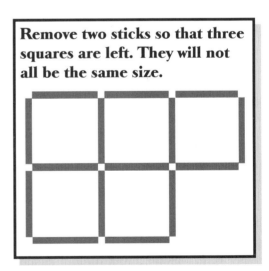

Remove two sticks so that three squares are left. They will not all be the same size.

Multiplication

How Many Colors?

How many different colors do you need to color a map? Sections that have the same color should not touch.

Copy these figures on your paper and try to color them. Use the least number of colors possible.

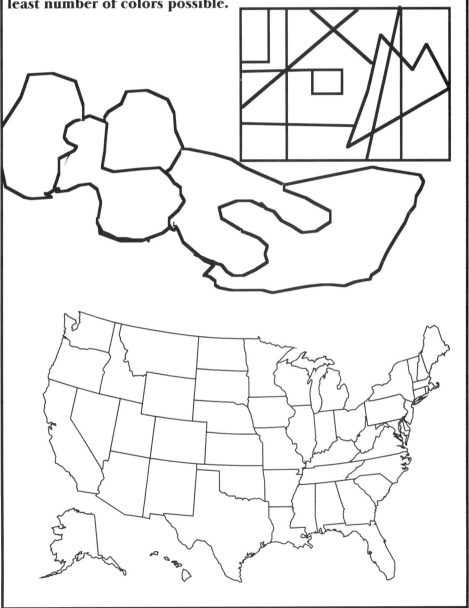

Multiplication

6.10 Units Story Problems

Many story problems ask you to change units. You have to remember the units we studied in Chapter 2 to do the problems. Some of these units are listed below.

Units Chart

1 foot (ft.)	=	12 inches (in.)
1 yard (yd.)	=	3 feet (ft.)
1 mile (mi.)	=	1,760 yards (yd.)
1 minute (min.)	=	60 seconds (sec.)
1 hour (hr.)	=	60 minutes (min.)
1 day	=	24 hours (hr.)
1 week (wk.)	=	7 days
1 pound (lb.)	=	16 ounces (oz.)
1 ton	=	2,000 pounds (lb.)
1 quart (qt.)	=	2 pints (pt.)
1 gallon (gal.)	=	4 quarts (qt.)
1 tablespoon (T.)	=	3 teaspoons (t.)

Model 2

Julie has five pounds of chocolates. How many ounces does she have?

The key words are pounds and ounces. The Units Chart says that 1 lb. = 16 oz. This is changing big units to small units. Pounds are bigger than ounces. The number to multiply with is 16. This is on the chart.

Person   **Numbers**   **Units & Key Words**

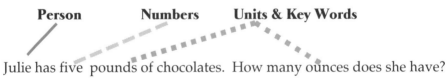

Julie has five pounds of chocolates. How many ounces does she have?

We can make many new problems by changing the person, numbers, and units.

Person
    Julie she   Pete he   Frank he

Numbers
    5   3   6

Units & Key Words
    Pounds of chocolate/oz.   gallons of milk/qt.   feet tall/inches

Multiplication

New Problem #1:
Pete has three gallons of milk. How many quarts does he have?
New Problem #2:
Frank is six feet tall. How many inches tall is he?

6.11 PRACTICE
Write two new problems. Use other persons, numbers, and units.

6.12 HOMEWORK
Read these problems. List the person, numbers, units, and information from the Units Chart. Then solve the problem.

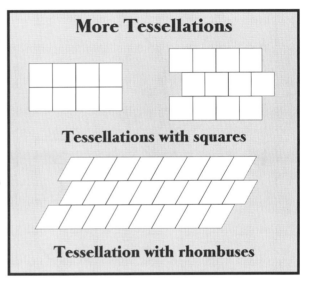

More Tessellations

Tessellations with squares

Tessellation with rhombuses

Example: Monica has eleven minutes left to do her test. How many seconds does she have?
Person: Monica
Numbers: 11
Units: minutes and seconds
From the chart: 1 minute = 60 seconds

```
   11
  x 60
  660  seconds to do her test
```

1. William has 100 yards to run. How many feet does he have to run?
2. Dwayne has three hours to drive. How many minutes does he have?
3. Hilda has two miles to ride home. How many yards does she have?
4. Charles has eight pounds of steak. How many ounces does he have?
5. Mohammed has twelve quarts of oil for his car. How many pints does he have?
6. Alice has thirty feet of muslin material. How many inches does she have?
7. Jason has four days until his birthday. How many hours does he have?

Multiplication

Dividing a Cube
Here are four ways to divide a cube into halves. Can you think of any others?

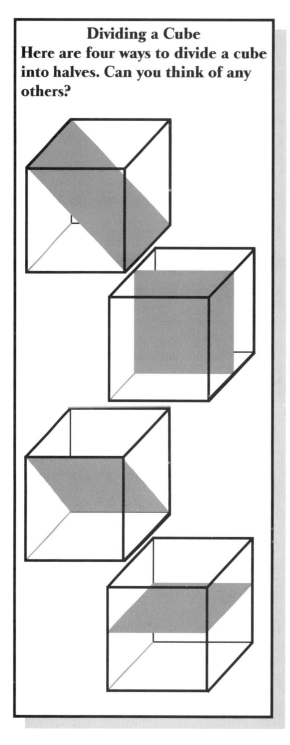

8. Eli has ten tons of sand. How many pounds does he have?

9. Sean has twelve days until the test. How many hours does he have?

10. Teresa has five tablespoons of sugar. How many teaspoons does she have?

11. Mark has seven pounds of flour. How many ounces does he have?

12. Jennifer has four gallons of milk. How many quarts does she have?

6.13 Vocabulary

Big units to small units
Bigger
Change
Left
Listed
Milk
Remember
Tall

Multiplication

6.14 Multiplying Decimal Fractions

If you can multiply, you can multiply decimal fractions. It looks the same and it works the same. It only has one more rule. The rule tells you where to put the decimal point in the answer.

```
    1.7          1.7          1.7          .17
  x 45         x 4.5         x .45        x .45
   76.5         7.65         .765        .0765
```

These examples are all the same except for the decimal points. Count how many digits are after the decimal point in both numbers.

```
    one          two         three         four
    1.7          1.7          1.7          .17
   x 45         x 4.5        x .45        x .45
```

Now count how many digits are after the decimal point in the answers.

```
   76.5         7.65         .765        .0765
    one          two         three         four
```

The rule for multiplying decimals fractions tells us to have the same number of digits after the decimal point in the **problem** and the **answer**.

6.15 Practice

Solve these problems.

1.   4
    x 2

2.   .5
    x .6

3.   7.1
    x .9

4.   7.1
    x .3

5.   .8
    x 2

6.   .17
    x .2

7.   2.6
    x .5

8.   .58
    x .4

9.   3.9
    x 3

10. Daniel has 15.6 boxes of ceramic tiles. Each one has 12 pieces of tile. How many tiles are in the boxes?
11. Raymond has 4.3 bags of sugar. Each one has 2.5 pounds of sugar. How many pounds of sugar are in the bags?

Multiplication

6.16 HOMEWORK

Set A: Solve these problems.

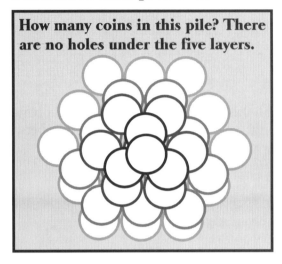

How many coins in this pile? There are no holes under the five layers.

1. 1.3
    x 2.2

2. 27
    x 49

3. 4.1
    x .10

4. 8.41
    x 3.8

5. 10.4
    x 4.19

6. 563
    x 8.3

7. .282
    x 75

8. 655.1
    x 5.1

9. 79.83
    x .61

10. 994.6
    x 1.07

11. 107.27
    x .923

12. 361.05
    x 6.8

Set B: Read these problems. List the person, numbers, units, and information from the Units Chart. Then solve the problem.

Example: Monica has 11.5 minutes left to do her test. How many seconds does she have?
Person: Monica
Numbers: 11.5
Units: minutes and seconds
From the chart: 1 minute = 60 seconds
   11.5
   x 60
   690.0 seconds to do her test

Multiplication

13. Francesco has 12.5 packages of trading cards. Each one has six cards. How many cards are in the packages?
14. Tanisha has 4.8 boxes of staples. Each one has 1,000 staples. How many staples are in the boxes?
15. Putri has 2.3 bags of cookies. Each one has eighty cookies. How many cookies are in the bags?
16. Tran has 3.2 pounds of candy. How many ounces does he have?
17. Ned has 6.9 pounds of flour. How many ounces does he have?
18. Frank ran 3.2 miles. How many yards did he run?
19. Carlos bought 8.7 gallons of gas. How many quarts did he buy?
20. Max has 8.5 cases of soda. Each one has twenty-four cans of soda. How many cans are in the cases?
21. Fabrice has 6.2 cans of coffee. Each one has thirteen ounces. How many ounces are in the cans?
22. Ophelia bought 3.5086 tons of gravel. How many pounds did she buy?

6.17 Vocabulary

Decimal fraction
Decimal point
Digits
Except
Rule

Which of these figures is impossible?

A.
side view
front view
top view

B.
side view
front view
top view

C.
side view
front view
top view

Multiplication

Binary Numbers

Binary numbers use the digits 0 and 1 only. Computers and calculators use these numbers to do their calculations.

0	=	0	25	=	11001
1	=	1	26	=	11010
2	=	10	27	=	11011
3	=	11	28	=	11100
4	=	100	29	=	11101
5	=	101	30	=	11110
6	=	110	31	=	11111
7	=	111	32	=	100000
8	=	1000	33	=	100001
9	=	1001	34	=	100010
10	=	1010	35	=	100011
11	=	1011	36	=	100100
12	=	1100	37	=	100101
13	=	1101	38	=	100110
14	=	1110	39	=	100111
15	=	1111	40	=	101000
16	=	10000	41	=	101001
17	=	10001	42	=	101010
18	=	10010	43	=	101011
19	=	10011	44	=	101100
20	=	10100	45	=	101101
21	=	10101	46	=	101110
22	=	10110	47	=	101111
23	=	10111	48	=	110000
24	=	11000			

6.18 Multiplication Story Problems with Money

We multiply money many times. We multiply money when we buy several things at the same price. We multiply money when we buy things by the pound. Look at these models.

Model 3:
Joe bought six gallons of ice cream at $1.97. How much did he spend?

Model 4:
William bought 3.2 lbs. of ham. It cost $3.69 per pound. How much did it cost?

Model 5:
Isabel wants to buy 19 pencils. They cost $.12 each. How much does she need?

The key words are **at**, **per**, and **each**. These words always tell you to multiply. Sometimes you have to round the answer like in Model 4.

Model 3:
Joe bought six gallons of ice cream at $1.97. How much did he spend?
Person: Joe
Numbers: 6 and $1.97
Units: gallons of ice cream and dollars
Key Words: at

$$\begin{array}{r} \$1.97 \\ \times\ \ 6 \\ \hline \$11.82 \end{array}$$

Multiplication

Making Tessellations

You can make your own custom tessellations by starting with a square or hexagon and following these steps.

Cut out a section from one side.

Add the same shape to the other side so that it sticks out.

Do the same to the other sides of your figures.

Model 4:

William bought 3.2 lbs. of ham. It cost $3.69 per pound. How much did it cost?

Person: William
Numbers: 3.2 and $3.69
Units: pounds of ham and dollars
Key Words: per

$$\begin{array}{r}\$3.69\\\times 3.2\\\hline\$11.808\end{array}$$ which becomes $11.81

Multiplication

Model 5:
Isabel wants to buy 19 pencils. They cost $.12 each. How much does she need?
Person: Isabel
Numbers: 19 and $.12
Units: pencils and dollars
Key Words: each

$.12
x 19
$2.28

6.19 PRACTICE
Write one new problem for each model. Use new persons, numbers, and units.

6.20 HOMEWORK
Read these problems. List the person, numbers, units, and information from the Units Chart. Then solve the problem.
Example: Manny bought three ice cream bars at $.39. How much did he spend?
Person: Manny
Numbers: 3 and $.39
Units: ice cream bars and dollars

$.39
x 3
$1.17

1. Pete bought four packages of lunchmeat at $.79. How much did he spend?
2. Nellie wants to buy six cans of soup. They cost $.58 each. How much does she need?

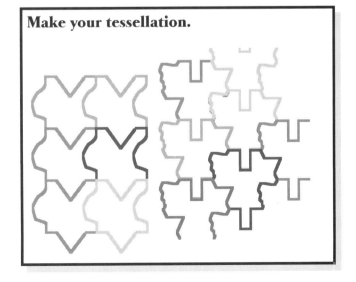

Make your tessellation.

151

Multiplication

3. Manuel bought nine candy bars at $.59. How much did he spend?
4. Francis bought three dozen eggs at $.99. How much did he spend?
5. Carolina bought 5.7 lbs. of apples. They cost $.89 per pound. How much did they cost?
6. Robert bought 39.8 feet of lumber. It cost $2.59 per foot. How much did it cost?
7. Alexia bought three balloons. They cost $1.99 per balloon. How much did they cost?
8. Tom wants to buy four boxes of nails. They cost $2.99 each. How much money does he need?
9. Luis wants to buy 16 quarts of oil. They cost $.79 each. How much money does he need?
10. Claude bought a bottle of soda at $1.68. How much did he spend?
11. Svetlana bought 1.75 lbs. of coffee. It cost $8.99 per pound. How much did she spend?
12. Doug wants to buy five pizzas. They cost $8.99 each. How much money does he need?

6.21 Vocabulary

At	Money
Buy	Pencils
By the pound	Per pound
Cost	Price
Each	Several
Ham	Spend
Ice cream	

6.22 Multiplication in Percent Problems

Most percent problems use multiplication to find the answers. It is like multiplying decimal fractions. The only thing new to remember is to change the percent into a decimal fraction by moving the decimal point to the left two places.

This table will help you change percents into decimals.

1%	becomes	.01
5%	becomes	.05
10%	becomes	.10
29%	becomes	.29
100%	becomes	1.00
125%	becomes	1.25

Multiplication

Model 6
Find 17% of 42.

17% becomes .17. Then multiply.

```
   42
x .17      Two decimal places in the answer
 7.14
```

Find 6 percent of 59.
6% becomes .06. Then multiply.

```
   59
x .06
 3.54
```

The key words are **percent of** or the symbol and word **% of**. These always tell you to change the percent into a decimal fraction and then multiply the numbers.

6.23 PRACTICE

Solve these percent problems.
1. 75% of 80
2. 18% of 90
3. 59 percent of 25
4. 65 percent of 200
5. 21% of 21
6. 85 percent of 900
7. 85% of $25.00
8. 8% of $12.50
9. 5 percent of $2,000
10. 75% of $125,000

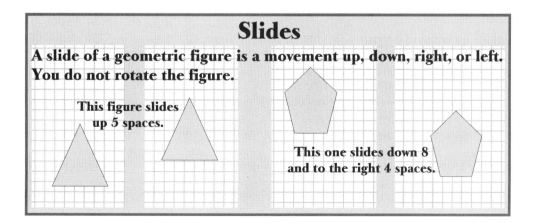

Slides

A slide of a geometric figure is a movement up, down, right, or left. You do not rotate the figure.

This figure slides up 5 spaces.

This one slides down 8 and to the right 4 spaces.

Multiplication

This is how you should have done the problems on page 153.

1. 75% of 80   .75 x 80 = 60
2. 18% of 90   .18 x 90 = 16.2
3. 59 percent of 25   .59 x 25 = 14.75
4. 65 percent of 200   .65 x 200 = 130
5. 21% of 21   .21 x 21 = 4.41
6. 85 percent of 900   .85 x 900 = 765
7. 85% of $25.00   .85 x $25.00 = $21.25
8. 8% of $12.50   .08 x $12.50 = $1.00
9. 5 percent of $2,000   .05 x $2,000 = $100
10. 75% of $125,000   .75 x $125,000 = $93,750

6.24 Homework

Solve these percent problems.

1. 55 percent of 99
2. 11% of 310
3. 62% of 849
4. 2 percent of 9,810
5. 74% of 1,671
6. 33% of 106
7. 87% of $718.00
8. 46 percent of 829
9. 9 percent of 144
10. 100 percent of $252
11. 28 percent of 53
12. 39% of 623
13. 11% of 4,156
14. 42 percent of 76
15. 61 percent of $195
16. 53% of 2,110
17. 74% of 323
18. 85 percent of 747
19. 96% of $8,545.00
20. 109% of $1,036.95
21. How much is 29% of $53,987.23?
22. How much is 15.5% of 90,000?
23. How much is 79.9% of 2,395?

6.25 Vocabulary

Change
Decimal fractions
Moving the decimal point
Percent
Percent of
Two places
% of

Multiplication

6.26 Using Formulas

We use many formulas in mathematics and science. Formulas are combinations of letters, numbers, and symbols. The letters are called variables. When two variables are written together we multiply.

Example: $A = lw$

This means to multiply l and w. If l is 10 and w is 5 we can evaluate the formula.

$$A = l \times w \quad \text{for } l = 10 \text{ and } w = 5$$
$$A = 10 \times 5$$
$$A = 50$$

If a formula has a number next to a variable we also multiply.

Example: $A = .5bh$ for b = 12 and h = 18

This mean to multiply .5, b, and h together. If b is 12 and h is 18 then we evaluate the formula this way.

$$A = .5 \times 12 \times 18$$
$$A = 6 \times 18$$
$$A = 108$$

Putting numbers into a formula is called substituting values for the variables. Finding the answer is called evaluating the formula.

Some formulas have other operations in them along with multiplication. Sometimes these operations are in parentheses, "( )." If a number or variable is next to a parentheses you multiply.

Example: $A = .5(b + c)h$ for b = 15, c = 21, and h = 7

This means that you first add what is inside the parentheses and then you multiply. Let's substitute 15, 21, and 7 for the variables and evaluate the formula.

$$A = .5(15 + 21)7$$
$$A = .5(36)7$$
$$A = 18 \times 7$$
$$A = 126$$

Multiplication

6.27 PRACTICE
Substitute the values for the variables and evaluate these formulas.
1. $A = lw$ $l = 21;\ w = 17$
2. $A = bh$ $b = 122;\ h = 240$
3. $A = .5bh$ $b = 25;\ h = 50$
4. $V = lwh$ $l = 6;\ w = 8;\ h = 20$
5. $M = DV$ $D = 0.25;\ V = 1.96$
6. $M = .03lf$ $l = 7;\ f = 1.236$
7. $A = .5(b_1 + b_2)h$ $b_1 = 6;\ b_2 = 12;\ h = 27$

6.28 HOMEWORK
Substitute the values for the variables and evaluate these formulas.
1. $A = lw$ $l = 1{,}200;\ w = 825$
2. $A = lw$ $l = 4{,}500;\ w = 698$
3. $A = bh$ $b = 127;\ h = 88$
4. $A = .5bh$ $b = 65;\ h = 87$
5. $A = .5bh$ $b = 124;\ h = 970$
6. $V = lwh$ $l = 14;\ w = 15;\ h = 25$
7. $V = lwh$ $l = 94;\ w = 17;\ h = 23$
8. $M = DV$ $D = 5.25;\ V = 0.855$
9. $M = .03lf$ $l = 72,\ f = 8.38$
10. $M = .03lf$ $l = 0.98;\ f = 0.45$
11. $A = .5(b_1 + b_2)h$ $b_1 = 10;\ b_2 = 82;\ h = 5$
12. $A = .5(b_1 + b_2)h$ $b_1 = 93;\ b_2 = 67;\ h = 82$
13. $V = .33lwh$ $l = 50;\ w = 60;\ h = 40$
14. $C = \pi d$ $\pi = 3.14159;\ d = 20$
15. $V = .5bhl$ $b = 7;\ h = 9;\ l = 25$
16. $F = 1.8C + 32$ $C = 36$
17. $D = .5att$ $a = 50;\ t = 10$
18. $y = 2x + 7$ $x = 15$
19. $x = .5y - 7$ $y = 16$

6.29 VOCABULARY

Combination	Operation
Evaluate	Parentheses
Formula	Science
Inside	Substitute
Letter	Symbol
Next to	Variable

Multiplication

6.30 MORE STORY PROBLEMS

These problems come from the lessons in this chapter. Look for the key words as you do the problems. The Units Chart will help you solve some of the problems.

6.31 HOMEWORK

Read these problems. List the person, numbers, units, key words, and information from the Units Chart. Then solve the problem.

1. Mary has three magazines. Each one has seventy-five pages. How many pages are in the magazines?
2. Sam wants to buy six hamburgers. They cost $1.29 each. How much money does he need?
3. How much is ten percent of $127.00?
4. Katrina bought twelve loaves of bread at $.89. How much did she spend?
5. Lomana bought seven gallons of gasoline. How many quarts did he buy?
6. How much is fifty percent of $96.40?
7. Willie wants to buy eleven candy bars. They cost $.89 each. How much does he need?
8. Jorge bought three bicycles at $129.99. How much did he spend?
9. How much is 25% of 480?
10. Pili has twelve bottles of soda. Each one has thirty-two ounces of soda. How many ounces are in the bottles?
11. Martha wants to buy fourteen sandwiches. They cost $1.79 each. How much does she need?
12. Julio worked six weeks at his new job. How many days did he work?
13. Ali bought 4.7 lbs. of bacon. It cost $1.97 per pound. How much did it cost?
14. Sourav bought 2.8 yards of concrete. It cost $18.50 per yard. How much did it cost?
15. Harold bought 12.9 gallons of gasoline. It cost $1.29 per gallon. How much did it cost?
16. Fernie bought 6.4 lbs. of cheese. It cost $2.89 per pound. How much did it cost?
17. Sara bought five pairs of skates at $49.99. How much did she spend?
18. Jim bought eighteen hotdogs at $1.19. How much did he spend?
19. Hannah bought 3.4 lbs. of steak. How many ounces did she buy?

Multiplication

20. Brad has nine boxes of candy. Each one has fifty-three pieces of candy. How many pieces of candy are in the boxes?
21. Gracie wants to buy two Christmas trees. They cost $15.00 each. How much does she need?
22. How much is 12% of 54?
23. Becky has fifteen rolls of tape. Each one has 125 feet of tape. How many feet of tape are on the rolls?
24. Andre has 15.5 ft. of string. How many inches does he have?

6.32 Chapter 6 Vocabulary Review

Use these words to answer the questions and problems below. Look up unfamiliar words in the dictionary. There might be more than one correct answer.

add	information into	property
answer	key words	putting
at	left	read
bag	letter	remember
become	list	round
big	many	rule
bigger	milk	same
book	model	science
bottle	move	several
box	moved	small
buy	money	solve
by	multiplication	spend
by the pound	multiply	story
change	next to	study
combination	number	substitute
commutative	operation	symbol
cost	other	tall
count	out loud	times
decimal fraction	page	together
decimal point	pair	two places
different	parentheses	units
digit	peanut	values
each	pencil	variables
equals	per pound	vertical
evaluate formula	percent of	vertically
except	per	where
ham	person	write
horizontal	place	written
horizontally	price	% of
ice cream	problem	

Multiplication

1. A problem can be written _____ or vertically.
2. The word _____ always means to multiply.
3. Multiplication is _____. This means that the numbers being multiplied can be moved.
4. Most story problems follow a _____.
5. The _____ _____ tell us what to do in a story problem.
6. Many story problems ask you to change _____.
7. You need the same number of _____ after the _____ in the problem and the answer of a multiplication problem.
8. We multiply _____ when we buy _____ things at the same price.
9. The key words at, _____, and _____ tell us to multiply in money problems.
10. Most percent problems use _____ to find the answer.
11. The letters in a formula are called _____.
12. When two variables are next to each other we _____.

6.33 Chapter Six Practice Test

Directions: Circle the correct answer.

1. 59 x 83 = 4,897, is written:
   a. multiplication
   b. vertically
   c. horizontally
   d. substitute

2. Multiplication is:
   a. several
   b. commutative
   c. parentheses
   d. combination

3. Which of these is a key word?
   a. each
   b. digit
   c. formula
   d. equals

4. Books and pages, bottles of soda and ounces, and bags of peanuts and peanuts are examples of:
   a. units
   b. equals
   c. values
   d. answers

5. Sam has four bags of marbles. Each one has 52 marbles. How many marbles are in the bags?
   a. 28 marbles
   b. 56 marbles
   c. 200 marbles
   d. 208 marbles

Multiplication

6. Juan has two pounds of candy. How many ounces does he have?
    a. 6 ounces
    b. 8 ounces
    c. 32 ounces
    d. 48 ounces

7. 4.7 x .28 =
    a. 1.316
    b. 13.16
    c. 131.6
    d. 1316

8. 4.71 x 82.4 =
    a. 3881.04
    b. 388.104
    c. 38.8104
    d. 3.88104

9. Joe bought eight gallons of milk at $2.66. How much did he spend?
    a. $2.74
    b. $10.66
    c. $21.28
    d. $21.58

10. Evaluate A = lw for l = 7 and w = 5.
    a. 2
    b. 12
    c. 30
    d. 35

11. How much is 20 percent of 82?
    a. 102
    b. 82.2
    c. 16.4
    d. 16

12. How much is 13.8% of 147?
    a. 20.286
    b. 147.138
    c. 160.8
    d. 202.86

6.34 Answers to Puzzles
From Section 6.3

From Section 6.4
These letters are more or less symmetrical depending on the way they are written.

| A | B | C | D | E | H | I | M |
| O | T | U | V | W | X | Y | |

Multiplication

From Section 6.9
Mathematicians claim that any map can be colored with only three different colors, even though map makers normally use five colors.

From Section 6.16
Bottom layer	19
Fourth layer	12
Third layer	7
Second layer	3
Top layer	1
TOTAL	42

From Section 6.17
Figure C is impossible.